Monotone Games

Tarun Sabarwal

Monotone Games

A Unified Approach to Games with Strategic Complements and Substitutes

Tarun Sabarwal
Department of Economics
The University of Kansas
Lawrence, KS, USA

ISBN 978-3-030-45515-6 ISBN 978-3-030-45513-2 (eBook)
https://doi.org/10.1007/978-3-030-45513-2

This Palgrave Pivot imprint is published by the registered company Springer Nature Switzerland AG
The registered company address is: Gewerbestrasse 11, 6330 Cham, Switzerland

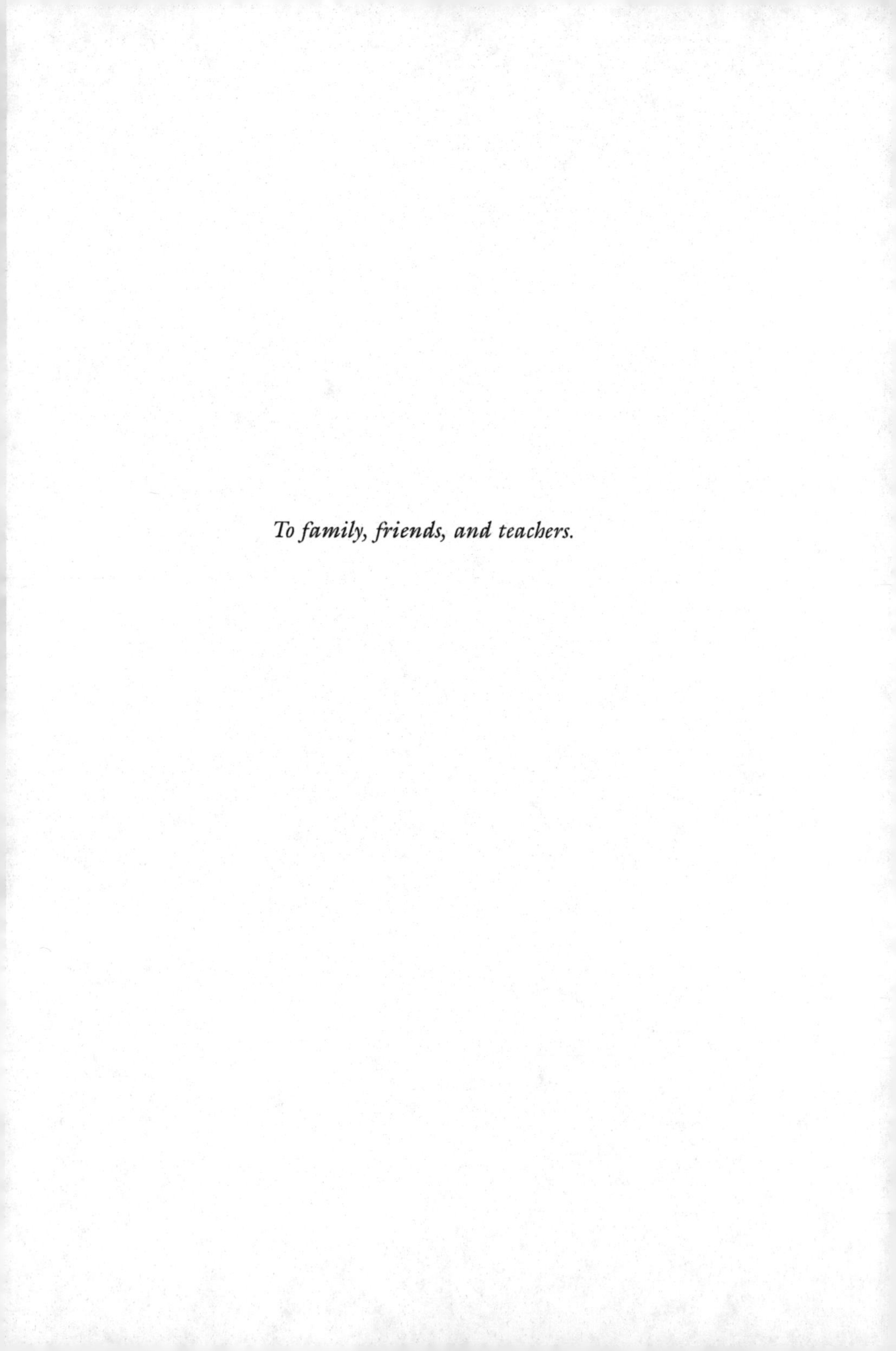

To family, friends, and teachers.

Preface

In many socioeconomic situations, decentralized decisions with interdependent effects may be studied using the framework of monotone games. This book develops the theory of monotone games in a manner that unifies the study of games with strategic complements, games with strategic substitutes, and combinations of the two.

Games with strategic complements have been studied for some time now and specialized cases of strategic substitutes are known as well. The study of general games with strategic substitutes and the study of general games with both strategic complements and substitutes are newer. The newer results are scattered in academic journals and unavailable in an integral form. Systematic connections across different classes of games are unavailable as well.

A goal here is to develop anew the foundations of all three classes of games in a unified manner under the umbrella of monotone games. In the process, existing results across different classes of games are proved anew under uniform assumptions and with a focus on unifying arguments, fundamental concepts are isolated and developed independently, new results are proved, new implications of existing results are derived, and many new examples and counterexamples are provided. Additional sections are dedicated to the underlying lattice theory and to develop a more cohesive treatment of codirectional and contradirectional incentives with many illustrative examples.

Another goal is to make the theory more accessible to others. There is a focus on development of more granular arguments that are easier to comprehend and can be combined in new ways to derive more general results for different cases. Proofs of results are expanded to include intermediate steps frequently skipped in academic journals, more consistent and streamlined notation is developed, and key arguments are repeated to make them transparent. Examples are reduced to their essentials to highlight the main points and to mention additional interpretations that may kindle more connections and applications.

There is a complementarity between these goals. A clear and unified development of the theory and its applications makes the material more accessible to a broader audience. Greater access increases research in this area and contributes to further development of our shared understanding of these ideas.

A motivating belief is that core principles studied here arise in fundamental ways in a large body of human and socioeconomic interaction with interdependent effects. There is a compelling reason for this body of knowledge to be accessible to a broader audience.

In an enterprise of this type, several decisions are made about the particular ways in which to corral and sift through the large amount of specialized knowledge available and to identify new connections. Along the way, additional decisions are made about what to include and what to exclude. Unification comes at the cost of excluding material less central to the main argument. Given the space and time constraints for this publication, the tradeoff is nontrivial.

Standing on the shoulders of others, I am fortunate to contribute in central ways to the foundations of the newer theory. If this work can help some to see further, perhaps a part of my debt to parents and teachers is repaid.

Lawrence, USA Tarun Sabarwal

Contents

List of Figures

List of Tables

CHAPTER 1

Introduction

Abstract Monotone games provide a natural framework to study decentralized decisions with interdependent effects in many socioeconomic situations. These include situations in which participants have an incentive to move in the same direction as others, an incentive to move in the direction opposite to others, or a combination of the two. Canonical examples help to identify the essential ideas behind each class of interactions and point toward the mathematical framework for a unified theory.

Keywords Coordination · Opposition · Strategic complements · Strategic substitutes · Monotone incentives

In economic and social environments, a decision by one participant has an effect on the well-being of others. When institutions, organizations, or individuals make decisions that affect the choices of others, the final results can often be complex and lead to unintended consequences.

Predictable patterns emerge when decisions are influenced in particular ways by the interdependent behavior of other decision- makers and by the decision-making environment itself. At the heart of this book is a study of principles governing environments in which participants have incentives to take decisions in particular directions, either decisions that are coordinated with those of others, or decisions that are opposed to those of others, or combinations of the two.

T. Sabarwal, *Monotone Games*,
https://doi.org/10.1007/978-3-030-45513-2_1

One class of decisions exhibits patterns of coordination, or movement in the same direction as other decision-makers. For example, if an increasing number of depositors make a run on a bank, it is in the best interest of other depositors to do the same before money runs out. This basic principle manifests in many seemingly unrelated socioeconomic situations such as speculative currency attacks, political uprisings, social network success or failure, run on groceries in a pandemic, online shopping platform success or failure, segregation and desegregation, propagation of health epidemics or misinformation on networks, coordinating successful social events, among others.

Environments in which decision-making is governed by incentives to move in the same direction as others are classified as strategic complements. A canonical example is a two player, two action coordination game.

Example 1.1 (Coordination game) Suppose there are two persons who are considering to adopt a technology to collaborate with each other. Each person may adopt either technology A or technology B. One person prefers technology A to technology B, the other prefers B to A. Choosing different technologies (either A by person 1 and B by person 2, or B by person 1 and A by person 2) is unhelpful for collaboration. In this situation, the well-being of each person depends not only on their own choice but also on choice of the other person and the incentive is for both people to coordinate, or move in the direction of choosing the same technology as the other person.

For ease of analysis, well-being of a person may be summarized by a numerical payoff to that person from the collective choice of both persons. For example, the situation here may be summarized in the form of a 2×2 bimatrix given in panel (1) of Table 1.1 (page 3). The first number in each cell is payoff for player 1 and the second number is payoff for player 2. When player 1 chooses A and player 2 chooses A, the first number, 2, in the corresponding cell is player 1's payoff from this collective choice (A, A) and the second number, 1, is player 2's payoff from this collective choice. Similarly, when collective choice is (B, B), player 1 has payoff 1 and player 2 has payoff 2, and when collective choice is either (A, B) or (B, A), each player has payoff 0.

Another class of decisions exhibits patterns of opposition, or movement in the direction opposite from that of others. For example, if two firms produce the same good and the output of firm 2 goes up, the profit maxi-

Table 1.1 Standard 2 × 2 Games

(1) Coordination

P1 \ P2	A	B
A	2, 1	0, 0
B	0, 0	1, 2

(2) Dove Hawk

P1 \ P2	D	H
D	2, 2	1, 3
H	3, 1	0, 0

(3) Matching Pennies

P1 \ P2	H	T
H	−1, 1	1, −1
T	1, −1	−1, 1

(4) Team Trust

P1 \ P2	C	D
C	3, 3	1, 4
D	4, 1	2, 2

mizing choice for firm 1 may be to produce less (because if it produces more as well, its profits are lower due to oversupply in the industry). This basic principle emerges in many different socioeconomic environments involving competition for a shared resource such as a public park, police services, river water use, roads and transportation, public education, limited funds for research grants, and so on. In each instance, the more intensively others use the shared resource, the best choice for a given participant is to use it less intensively, because their additional benefit (or probability of success) from more intensive use decreases when others use the resource more intensively. For example, if more motorists drive on a particular road, a given motorist is more likely to move to a different road, or if crime in a neighborhood is increasing, a resident is more likely to move to a different neighborhood.

Environments in which decision-making is governed by incentives to move in a direction opposite to others are classified as strategic substitutes. A canonical example is a two player, two action dove hawk game.

Example 1.2 (Dove Hawk) Suppose two persons are competing to share an available resource. Each person can compete less aggressively for the resource (summarized as playing a dove action, D) or compete more aggressively (a hawk action, H). If both compete less aggressively, they can share the resource with low cost of competition and their well-being is high. If both compete more aggressively, the cost of competition is high and the well-being of each suffers. If one person competes more aggressively and the other less aggressively, the more aggressive person dominates the use of resource and is better off as compared to the less aggressive person.

In this situation, if one person competes less aggressively, the incentive for the other person is to compete more aggressively, that is, to move in the direction opposite to that of the other person. This situation may be summarized with payoffs given in the 2×2 bimatrix in panel (2) of Table 1.1 (page 3).

A third class of decisions exhibits patterns of both coordination and opposition. For example, in law enforcement, police want to be at the same place as a criminal (coordination move), but criminal wants to be in a place different from police (opposing move). The same dynamic is present between advertisers and consumers, a baseball pitcher and a hitter, hunter and hunted, dictator and rebel, and so on. This class of decisions is made more complex when there are subsets of decision-makers with codirectional incentives for coordination and others with contradirectional incentives for opposing moves. A canonical example is a two player, two action game of matching pennies.

Example 1.3 (Matching Pennies) Suppose two persons each have a penny. Each person chooses a side of the penny to play (heads, H, or tails, T). When their choices are revealed, if their choices match then the first person gives the second person one dollar, otherwise, the second person gives one dollar to the first. In this situation, incentive for player 2 is to match or move in the same direction as player 1, and incentive for player 1 is to mismatch or move in the direction opposite to player 2. Payoffs are given in the 2×2 bimatrix in panel (3) of Table 1.1 (page 3).

Another class of situations is when there is tension between individual interest and group interest. For example, a cartel has a collective interest to limit output to increase their collective profit, but the individual interest of each member of the cartel is to produce more than their quota. Or, it may be in a country's best interest to limit activity to curtail the spread of disease in a pandemic, but it may be in each small firm's interest to remain open so it does not fail. This principle manifests in situations where individual interest is in conflict with a higher group interest, which can only be achieved by a level of trust or cooperation that is absent without additional incentives or enforcement. A canonical example is a two player, two action team trust game. This may also be viewed as prisoners' dilemma.

Example 1.4 (Team trust game) Suppose two persons are working on a team project and share the project reward. Each person can cooperate and put in high effort (C) or defect and put in low effort (D). If both cooperate and put in high effort, the project has high success and both members of the team are well off. If both put in low effort, the project has low success and each receives low payoff. If one person puts high effort and the other puts low effort, the project has moderate success. In this case, the person who puts in less effort shares the reward without incurring the cost of high effort and receives a high final payoff, and the person who puts in high effort shares the reward but has a high cost of effort to receive a low final payoff. This may be summarized with payoffs given in the 2×2 bimatrix in panel (4) of Table 1.1 (page 3).

Collectively, these are situations in which participants face *monotone interdependent incentives*, or more briefly, *monotone incentives*. *Monotone games* provide a framework to study the pattern of rational and decentralized decisions by participants facing monotone incentives and their collective equilibrium impact when all participants are simultaneously trying to do the best for themselves and there is no tendency for change. This turns out to apply to a large body of economic and social interactions.

In order to study movement in different directions more generally, the notion of direction has to be defined. In many contexts this may be natural. For example, in the coordination game (panel (1) of Table 1.1, page 3), if there is some information that technology A is worse than technology B, we may say A is lower than B, or B is higher than A, denoted $A \prec B$. On the other hand, if technology B is worse than technology A, we may say B is lower than A, or A is higher than B, denoted $B \prec A$. Moving in the same direction would mean that if one player takes a higher action, then it is in the other player's best interest to take a higher action as well.

As this shows, there may be more than one way to define direction. Sometimes the description of the game provides an intrinsic definition. Other times, the analyst describing the game may provide a reasonable definition.

A general tool to formalize direction is partial order on the set of actions. In the coordination game, the action set of each player is $\{A, B\}$ and one partial order on this set is $A \prec B$. Another partial order is $B \prec A$.

Direction may be defined using a natural partial order on the set of actions. Direction or partial order on the set of actions is viewed as a primitive of the environment in which joint interaction is studied.

Given an action set for each player, the collection of all joint actions by the players is the Cartesian product of these sets. An element of the Cartesian product is a profile of actions, which includes one action for each player. The idea of direction for the collection of joint actions is formalized by the product partial order on this Cartesian product of action sets. In the coordination game, the Cartesian product of action sets of the two players is $\{(A, A), (A, B), (B, A), (B, B)\}$.

The well-being of a player is summarized by a real number that depends on the profile of actions, which includes an action for each player. This allows for interdependent incentives. In other words, the well-being of a player is summarized by a real-valued function on the Cartesian product of the action sets. In the coordination game, the payoff for player 1, denoted u_1, is given by $u_1(A, A) = 2$, $u_1(A, B) = u_1(B, A) = 0$, and $u_1(B, B) = 1$. The payoff for player 2, denoted u_2, is given by $u_1(A, A) = 1$, $u_1(A, B) = u_1(B, A) = 0$, and $u_1(B, B) = 2$.

With this formulation, when player 2 takes action A, if player 1 takes action A as well, player 1 payoff is 2, and if player 1 takes action B, player 1 payoff is 0. Therefore, when player 2 takes action A, taking action A is in the best interest of player 1. Similarly, if player 2 takes action B, it is in player 1's best interest to coordinate and take action B. Moreover, when player 2 moves to take a higher action (from A to B), player 1 moves in the same direction and takes a higher action. In other words, in this environment, player 1 has strategic complements. Similarly, player 2 has strategic complements. This is a game with strategic complements.

The incentives are reversed in the dove hawk game. With $D \prec H$, when player 2 takes a higher action (goes from D to H), it is in player 1's best interest to take a lower action (goes from H to D). With $H \prec D$, if player 2 takes a higher action (goes from H to D), it is in player 1's best interest to take a lower action (goes from D to H). In other words, with either partial order on $\{D, H\}$, player 1 has strategic substitutes. Similarly, player 2 has strategic substitutes. This is a game with strategic substitutes. The case of changing the order on one player's action space and not on the other is discussed in Chapter 4 (page 79).

In matching pennies, both incentives are present. With $H \prec T$, player 1 has strategic substitutes, because their incentive is to move in the direction opposite to player 2 choice, and player 2 has strategic complements, because their incentive is to move in the same direction as player 1 choice. This is a game with both strategic complements and strategic substitutes.

In the team trust game, player 1 incentive is to play D regardless of what player 2 chooses. The same holds for player 2.

All these are examples of monotone games. Consider the following additional examples.

Example 1.5 (Multi-player coordination game) Suppose there are many people in a society and each person can take one of two actions, a baseline action, denoted 0, or a new action, denoted 1. Each person has an incentive to coordinate their action with others in the sense that when a greater number of people in society play an action, each person has a greater incentive to play the same action. Playing the baseline action gives a person a baseline payoff normalized to 0, regardless of the actions of others. Payoff from action 1 depends positively on the fraction of people in society who play 1 and negatively on the fraction of people who play 0.

This can be formalized as the following game. There are finitely many players I, indexed $i = 1, \ldots, I$. Each player i can take action $x_i \in \{0, 1\}$. A profile of actions is a collection of actions, one for each player, denoted $x = (x_1, \ldots, x_I)$. For convenience, we may write this in partition form (x_i, x_{-i}) where x_i is action of player i and x_{-i} is the profile of actions of all players other than i. Player i payoff depends on actions of all players as follows. For profile of other player actions x_{-i}, the payoff to player i from playing 0 is $u_i(0, x_{-i}) = 0$, and the payoff to player i from playing 1 is

$$u_i(1, x_{-i}) = 1\left(\frac{1}{I-1}\left(\sum_{j \neq i} x_j\right)\right) - 1\left(1 - \left(\frac{1}{I-1}\left(\sum_{j \neq i} x_j\right)\right)\right).$$

In other words, if player i plays 1, they receive a payoff between -1 and 1 depending on the fraction of others who play 1, given by $\frac{1}{I-1}(\sum_{j \neq i} x_j)$. Payoff from playing 1 is increasing in the fraction of others who play 1 and decreasing in the fraction of others who play 0.

Example 1.6 (Multi-player externality game) Consider a situation in which there is a group of people and there is a good the production of which benefits everyone in the group. For example, once an open source software (operating system, text editing, web design, and so on) is produced, all users benefit from it. Once an open access article is published, all researchers benefit from it. Once a free online teaching platform is established, all users benefit from it. Once a social media product (meme, gif, video, tutorial, and so on) is produced and distributed, all recipients can use it. Once a public broadcast is produced and distributed, all viewers benefit from it.

For convenience, the benefit of the good to each person is normalized to 1 and the good can be produced by a single person by incurring a positive cost $c < 1$. Each person can either produce the good (take action 1) or not (action 0). Once produced, there is no significant additional benefit for anyone else to produce the same good.

This can be formalized as the following game. There are finitely many players I, indexed $i = 1, \ldots, I$. Each player i can take action $x_i \in \{0, 1\}$. A profile of actions is a collection of actions, one for each player, denoted $x = (x_1, \ldots, x_I)$ or $x = (x_i, x_{-i})$. For profile of other player actions x_{-i}, if player i chooses to produce the good, payoff to player i is $u_i(1, x_{-i}) = 1 - c$. If player i chooses not to produce the good, payoff is

$$u_i(0, x_{-i}) = \begin{cases} 1 \text{ if } \sum_{j \neq i} x_j > 0, \\ 0 \text{ if } \sum_{j \neq i} x_j = 0. \end{cases}$$

Example 1.7 (Bertrand oligopoly) Consider an industry in which a group of firms produces similar but differentiated products. For example, the industry of branded handbags, or bottles of perfume, or branded cereal, or Chinese food restaurants, or running shoes, and so on. Each firm can choose the price at which to sell its product. Demand for each firm's product depends inversely on the price it charges and depends positively on the price other firms charge for their product. Each firm can produce the good using an easily available constant returns to scale technology.

This can be formalized as the following game. There are finitely many firms I, indexed $i = 1, \ldots, I$. Each firm i can choose its price per unit $x_i \in [0, x^{\max}] \subset \mathbb{R}$, where $x^{\max}$ is maximum unit price possible given demand for the firm's product. Given a profile of firm prices $x = (x_1, \ldots, x_I)$, or $x = (x_i, x_{-i})$, quantity of market demand for firm i product (denoted q_i) depends inversely on the firm's own price and positively on price of other firms and is given by $q_i = (a - bx_i + \beta(\sum_{j \neq i} x_j))$, where a, b, and β are positive parameters summarizing demand conditions in the industry, with $\beta(I - 1) < 2b$ (to ensure positive price in equilibrium). Marginal cost of production is given by a nonnegative number c, with $c < a$. For a profile of other firm prices x_{-i}, profit to firm i from choosing price x_i is total revenue minus total cost, that is,

$$u_i(x_i, x_{-i}) = q_i x_i - cq_i,$$

where $q_i x_i = (a - bx_i + \beta(\sum_{j \neq i} x_j))x_i$ is total revenue to firm i from selling quantity q_i units of output at price x_i per unit, and cq_i is total cost of producing quantity q_i units.

Example 1.8 (Cournot oligopoly) Consider an industry in which a group of firms produces a homogeneous product. For example, the industry producing navel oranges, or retail gasoline, or reduced fat milk, or corn, or oil changes, or wheat, or strawberries, and so on. The price of the product depends inversely on total output being sold in the industry and is determined by demand conditions in the market. Each firm can produce the good using a standard existing constant returns to scale technology.

This can be formalized as the following game. There are finitely many firms I, indexed $i = 1, \ldots, I$. Each firm i can produce output $x_i \in [0, x^{\max}] \subset \mathbb{R}$, where $x^{\max}$ is maximum output possible with existing firm capacity. Given a profile of firm outputs $x = (x_1, \ldots, x_I)$, or $x = (x_i, x_{-i})$, inverse market demand is given by $p = (a - b(\sum_{i=1}^{I} x_i))$, where a and b are positive parameters summarizing demand conditions in the industry. Marginal cost of production is given by a nonnegative number c, with $c < a$. For a profile of other firm outputs x_{-i}, profit to firm i from choosing output x_i is total revenue minus total cost, that is,

$$u_i(x_i, x_{-i}) = \left(a - b\left(\sum_{i=1}^{I} x_i\right)\right) x_i - cx_i,$$

where $\left(a - b\left(\sum_{i=1}^{I} x_i\right)\right) x_i$ is total revenue to firm i from selling x_i units, and cx_i is total cost of producing x_i units.

In each of the last four examples, there is a natural order on the actions of each player. This can be extended to a partial order on the set of profiles of actions, and combined with the structure of payoffs, it can be deduced that each player has either strategic complements or strategic substitutes.

These examples show some of the versatility of monotone interdependent incentives to model a variety of phenomena. These are canonical models and their generalizations and modifications are used in many real world applications.

The coordination game (Example 1.1, page 2), multi-player coordination game (Example 1.5, page 7), and Bertrand oligopoly (Example 1.7, page 8) are games with strategic complements. In each of these games, payoff for each player has codirectional incentives, that is, each player's payoff provides the player an incentive to move in the same direction as others.

The dove hawk game (Example 1.2, page 3), multi-player externality game (Example 1.6, page 7), and Cournot oligopoly (Example 1.8, page 9) are games with strategic substitutes. In each of these games, payoff for each player has contradirectional incentives, that is, each player's payoff provides the player an incentive to move in the direction opposite to others.

Matching pennies (Example 1.3, page 4) has both types of incentives. Player 2 has codirectional incentives (wants to move in the same direction as player 2) and player 1 has contradirectional incentives (wants to move in the direction opposite to player 2). Team trust game (Example 1.4, page 5) has both incentives as well, but in a degenerate sense.

A goal in this book is to isolate the underlying unifying principles governing these types of interactions, to understand similarities and differences in the structure of incentives that predict whether a player will move in the same direction as others or in the direction opposite to others, and to study the impact of these incentives on predicted outcomes when everyone is simultaneously trying to do the best for themselves and no one has an incentive to change their decision given the decisions of others.

In order to do that, the next chapter presents elements of the theory of partial orders and lattices, and the general framework of a lattice game. Using this framework, monotone games with strategic complements, monotone games with strategic substitutes, and monotone games with both strategic complements and strategic substitutes are studied in more detail in later chapters.

CHAPTER 2

Elements of Lattice Games

Abstract Partial order and lattice formalize the idea of direction of movement. Correspondence and fixed point provide the mathematical underpinning for best choices and equilibrium. A lattice game is a strategic game in which players have lattice action spaces. Foundations of monotone games are set in this framework. Concepts from the theory of normal form games provide the tools to analyze and predict outcomes in lattice games. The theory of lattice games develops these ideas and a diversity of examples highlights their applications.

Keywords Partial order · Lattice · Nash equilibrium · Dominance solvable · Globally stable · Lattice game

Chapter 1 points out elements of the mathematical structure useful to study monotone interdependent interactions and equilibrium outcomes emerging from these interactions.

This chapter presents some of the underlying mathematical framework useful throughout the book. Important components of this framework include the notions of partial order and lattice. These are used to formulate several ideas helpful for the analysis, including interval in a lattice, complete lattice, lattice set order, correspondence, and the set of fixed points of a correspondence.

T. Sabarwal, *Monotone Games*,
https://doi.org/10.1007/978-3-030-45513-2_2

In addition to this mathematical structure, this chapter develops the framework of lattice games. The class of lattice games is a general construct that naturally subsumes games with strategic complements, games with strategic substitutes, and combinations of the two. It provides a foundation for the study of these three classes that comprise monotone games. It includes additional cases as well. Several results for monotone games can be seen more generally at this level.

2.1 Lattices

A general tool to formalize the idea of direction is partial order on a set. A partially ordered set in which every pair of elements has a (smallest) larger element and a (largest) smaller element is a lattice. Standard material on lattices may be found in Birkhoff (1995) and in Topkis (1998). Results that are unavailable or hard to find elsewhere in the form used here are proved.

2.1.1 Partially Ordered Set

A binary relation $\preceq$ on a set X is reflexive, if for every $x \in X$, $x \preceq x$, it is antisymmetric, if for every $x, y \in X$, $x \preceq y$ and $y \preceq x \Rightarrow x = y$, and it is transitive, if for every $x, y, z \in X$, $x \preceq y$ and $y \preceq z \Rightarrow x \preceq z$. A ***partial order*** on a set X is a binary relation $\preceq$ that is reflexive, antisymmetric, and transitive. A ***partially ordered set***, or ***poset***, is a set X along with a partial order $\preceq$ on the set, denoted $(X, \preceq)$.

For a poset $(X, \preceq)$ and subset A of X, the ***relative partial order*** on A is defined as follows. For every $x, x' \in A$, $x \preceq_A x' \Leftrightarrow x \preceq x'$. It follows immediately from the definition that $(A, \preceq_A)$ is a poset in the relative partial order. For posets $(X, \preceq_X)$ and $(Y, \preceq_Y)$, the Cartesian product $X \times Y$ is a poset under the ***product partial order*** given by $(x, y) \preceq (x', y') \Leftrightarrow x \preceq_X x'$ and $y \preceq_Y y'$.

Examples 2.1 Consider the following examples.

1. Real numbers, $\mathbb{R}$, with their natural order $\leq$ form a poset $(\mathbb{R}, \leq)$. Its standard subsets $\mathbb{N}, \mathbb{Z}, \mathbb{Q}$ are all posets in the relative partial order. Any finite set of real numbers $\{x_1, ..., x_N\}$ is a poset in the relative partial order.
2. Finite-dimensional Euclidean space over the reals, $\mathbb{R}^N$, is a poset in the product partial order (also termed Euclidean partial order). For every $x = (x_1, ..., x_N)$ and $y = (y_1, ..., y_N)$ in $\mathbb{R}^N$, $x \preceq y \Leftrightarrow x_1 \leq y_1, x_2 \leq y_2, ..., x_N \leq y_N$.

3. The four point subset $\{(0, 0), (1, 0), (0, 1), (1, 1)\} \subset \mathbb{R}^2$ is a poset in the relative partial order from $\mathbb{R}^2$.
4. Real-valued functions on a set X, denoted $\mathbb{R}^X = \{f | f : X \to \mathbb{R}\}$ is a poset with the product partial order, which is also the pointwise partial order, given by $f \preceq g \Leftrightarrow \forall x \in X, f(x) \leq g(x)$. With appropriate assumptions on X, natural subsets of $\mathbb{R}^X$ such as continuous functions on X, or differentiable functions on X, or smooth functions on X, and so on are posets in the relative partial order. Similarly, when $X = \mathbb{N}$, the set of sequences of real numbers is a poset, as are subsets of ℓ_p sequences.
5. The power set of a set X, denoted $\mathcal{P}(X)$, is a poset under set inclusion. For every $A, B \subset X, A \preceq B \Leftrightarrow A \subset B$. In a topological space, natural subsets of $\mathcal{P}(X)$ given by open sets, or closed sets, or compact sets, and so on, are all posets in the relative partial order. In a measure space, measurable sets are a poset, as are sets of measure zero, and so on.
6. Consider an arbitrary two player game where X_1 and X_2 are action spaces for player 1 and 2, respectively, and $u_1 : X_1 \times X_2 \to \mathbb{R}$ and $u_2 : X_1 \times X_2 \to \mathbb{R}$ are their corresponding payoff functions. If $(X_1, \preceq_1)$ and $(X_2, \preceq_2)$ are posets, then the set of profiles of actions given by the Cartesian product $X_1 \times X_2$ is a poset in the product partial order: $(x_1, x_2) \preceq (x_1', x_2') \Leftrightarrow x_1 \preceq_1 x_1'$ and $x_2 \preceq_2 x_2'$.

Two points x, y in a poset $(X, \preceq)$ are ***comparable*** (or ***ordered***), if $x \preceq y$ or $y \preceq x$. In this case, we say that x is lower than y when $x \preceq y$, or x is higher than y when $y \preceq x$. Points x, y are ***strictly comparable*** (or ***strictly ordered***), if they are comparable and $x \neq y$. In this case, we say x is strictly lower than y, denoted $x \prec y$, or x is strictly higher than y, denoted $y \prec x$, as the case may be.

Two points x, y are ***incomparable*** (or ***unordered***), if they are not comparable, that is, $x \npreceq y$ and $y \npreceq x$. A partial order is ***complete*** (or ***linear***), if every pair of points is comparable. A poset with a complete order is a ***chain***. In other words, a chain is a poset in which every pair of points comparable. On the other hand, a poset in which every pair of points is incomparable is ***totally unordered***.

It is easy to check the following. If $(X, \preceq)$ is a chain, then so is every subset of X in the relative partial order. Moreover, if $(X, \preceq_X)$ and $(Y, \preceq_Y)$ are two posets, each with two elements that are strictly comparable, then

$X \times Y$ with the product partial order has incomparable elements, and is not a chain.

Examples 2.2 Consider the following examples.

1. The set of real numbers $(\mathbb{R}, \leq)$ is a chain, as are its subsets.
2. $(\mathbb{R}^N, \preceq)$, for $N \geq 2$, is a poset that is not a chain. Any two standard basis vectors are unordered. The subset consisting of standard basis vectors is totally unordered.
3. The four point subset $\{(0, 0), (1, 0), (0, 1), (1, 1)\} \subset \mathbb{R}^2$ is a poset that satisfies $(0, 0) \prec (1, 0) \prec (1, 1)$, and $(0, 0) \prec (0, 1) \prec (1, 1)$, and $(1, 0)$ and $(0, 1)$ are incomparable. Its subset $\{(1, 0), (0, 1)\}$ is totally unordered.
4. In the set of real-valued functions $(\mathbb{R}^X, \preceq)$ with pointwise partial order, a function f is lower than function g, $f \preceq g$, if for every $x \in X$, $f(x) \leq g(x)$. Graphically, f lies below g everywhere. Functions that cross each other are unordered.
5. In the power set $\mathcal{P}(X)$ partially ordered with set inclusion, subsets A and B are comparable, if, and only if, one is contained in the other. Subsets A and B are unordered, if, and only if, both their set differences are nonempty $A \setminus B \neq \emptyset$ and $B \setminus A \neq \emptyset$.
6. In every two player strategic game, $((X_1, u_1), (X_2, u_2))$, with partially ordered actions sets $(X_1, \preceq_1)$ and $(X_2, \preceq_2)$, if each action set has a pair of strictly comparable actions, then the space of profiles of actions, $X_1 \times X_2$, is a poset that is not a chain in the product partial order.

2.1.2 *Join and Meet*

Let A be a nonempty subset of a poset $(X, \preceq)$. An ***upper bound of*** A is an element $x \in X$ such that for every $a \in A$, $a \preceq x$. The ***join of*** A (or ***sup of*** A) is a least upper bound of A, that is, an element $x \in X$ such that x is an upper bound of A and for every upper bound x' of A, $x \preceq x'$. Antisymmetry of partial order implies that a least upper bound is unique, if it exists, and in this case, it is denoted $\bigvee A$ or $\sup A$. For two points $x, y \in X$, the ***join of*** x ***and*** y (or ***sup of*** x ***and*** y) is the sup of the two point set $\{x, y\}$ and is denoted $x \vee y$.

A ***lower bound of*** A is an element $x \in X$ such that for every $a \in A$, $x \preceq a$. The ***meet of*** A (or ***inf of*** A) is a greatest lower bound of A, that is, an element $x \in X$ such that x is a lower bound of A and for every lower bound x' of A, $x' \preceq x$. Antisymmetry of partial order implies that a greatest lower bound is unique, if it exists, and in this case, it is denoted $\bigwedge A$ or $\inf A$. For two points $x, y \in X$, the ***meet of x and y*** (or ***inf of x and y***) is the inf of the two point set $\{x, y\}$ and is denoted $x \wedge y$.

Examples 2.3 Consider the following examples.

1. In a chain X, the operations join and meet of two points are trivial. For every $x, y \in X$,

$$x \vee y = \begin{cases} x & \text{if } y \preceq x \\ y & \text{if } x \preceq y, \end{cases} \quad \text{and } x \wedge y = \begin{cases} y & \text{if } y \preceq x \\ x & \text{if } x \preceq y. \end{cases}$$

   In other words, $x \wedge y = \min\{x, y\}$ and $x \vee y = \max\{x, y\}$. If $X = \mathbb{R}$, and $A = \{1, 2, ..., 10\}$, then $\sup A = \bigvee A = 10$ and $\inf A = \bigwedge A = 1$. If $A = (0, 1]$, then $\sup A = \bigvee A = 1$ and $\inf A = \bigwedge A = 0$. If $A = \mathbb{N}$, then neither $\sup A$ nor $\inf A$ exist in $\mathbb{R}$.
2. In $\mathbb{R}^N$, for $x = (x_1, \ldots, x_N)$ and $y = (y_1, \ldots, y_N)$,

$$x \vee y = (\max\{x_1, y_1\}, \ldots, \max\{x_N, y_N\}) = (x_1 \vee y_1, \ldots, x_N \vee y_N)$$

   and

$$x \wedge y = (\min\{x_1, y_1\}, \ldots, \min\{x_N, y_N\}) = (x_1 \wedge y_1, \ldots, x_N \wedge y_N).$$

   More generally, suppose $A = \{(a_1^i, a_2^i, \ldots, a_N^i) \in \mathbb{R}^N \mid i \in I\}$, where I is a nonempty index set. Then

$$\bigvee A = \left(\bigvee_{i \in I} a_1^i, \bigvee_{i \in I} a_2^i, \ldots, \bigvee_{i \in I} a_N^i\right) \text{ and}$$

$$\bigwedge A = \left(\bigwedge_{i \in I} a_1^i, \bigwedge_{i \in I} a_2^i, \ldots, \bigwedge_{i \in I} a_N^i\right),$$

   assuming each component operation is well-defined.
3. In $\mathbb{R}^2$, consider

$$A = \{(0, 0), (1, 0), (2, 0), (0, 1), (1, 1), (2, 1), (0, 2), (1, 2), (2, 2)\},$$
$$B = \{(2, 0), (1, 1), (0, 2)\},$$
$$C = \{(0, 0), (1, 1), (2, 2)\}, \text{ and}$$
$$D = \{(0, 1), (1, 1), (2, 0)\}.$$

Then $\sup A = \sup B = \sup C = (2, 2)$, and $\inf A = \inf B = \inf C = (0, 0)$. Moreover, $\sup D = (2, 1)$ and $\inf D = (0, 0)$. Furthermore, B is totally unordered and C is a chain.

4. In $(\mathbb{R}^X, \preceq)$ with pointwise partial order, for functions $f, g \in \mathbb{R}^X$, $f \vee g$ is the function given by $(f \vee g)(x) = f(x) \vee g(x) = \max\{f(x), g(x)\}$ and $f \wedge g$ is the function given by $(f \wedge g)(x) = f(x) \wedge g(x) = \min\{f(x), g(x)\}$. More generally, suppose $A = \{f_i | i \in I\}$, where I is a nonempty index set. Then $\bigvee A$ is the function $\bar{f}$ given by $\bar{f}(x) = \sup_{i \in I} f_i(x)$ (assuming this is well-defined), and $\bigwedge A$ is the function $\underline{f}$ given by $\underline{f}(x) = \inf_{i \in I} f_i(x)$ (assuming this is well-defined). Graphically, these are, respectively, the upper and lower envelopes of the functions in A.
5. In the power set $\mathcal{P}(X)$, for subsets A and B of X, $A \vee B = A \cup B$ and $A \wedge B = A \cap B$. For a collection of subsets $\mathcal{A} = \{A_i \subset X | i \in I\}$, where I is a nonempty index set, $\bigvee \mathcal{A} = \bigcup_{i \in I} A_i$ and $\bigwedge \mathcal{A} = \bigcap_{i \in I} A_i$.

These examples show that the join or meet of a subset A may or may not exist. If it exists, it may or may not be in A. Moreover, different sets may have the same join or meet.

Join or meet of a subset A depends on the set of which A is a subset. For example, consider $X = \mathbb{R}^2$, $Y = \{(-1, 0), (1, 0), (0, 1), (2, 2)\}$, and $A = \{(1, 0), (0, 1)\}$. Then $A \subset X$ implies $\inf A = (0, 0)$ and $\sup A = (1, 1)$, but $A \subset Y$ implies $\inf A = (-1, 0)$ and $\sup A = (2, 2)$. When necessary, this is distinguished by including the superset in the notation. That is, $\inf_X A = (0, 0)$, $\sup_X A = (1, 1)$, $\inf_Y A = (-1, 0)$, and $\sup_Y A = (2, 2)$.

Proposition 2.4 *For every poset X and every $B \subset A \subset X$ with B nonempty, $\inf_A B \preceq \inf_X B \preceq \sup_X B \preceq \sup_A B$, whenever these are well-defined.*

Proof As every lower bound of B in A is a lower bound of B in X, it follows that $\inf_A B \preceq \inf_X B$, and as every upper bound of B in A is an upper bound of B in X, it follows that $\sup_X B \preceq \sup_A B$. Moreover, $\inf_X B \preceq \sup_X B$ follows from definition of inf and sup. □

2.1.3 *Lattice*

A ***lattice*** is a poset $(X, \preceq)$ in which for every $x, y \in X, x \wedge y \in X$ and $x \vee y \in X$. In other words, a lattice is a poset that is closed under meet and join. A subset A of a lattice X is a ***sublattice*** of X, if for every $x, y \in A$, $x \wedge y \in A$ and $x \vee y \in A$. In this definition, the meet and join operations are defined in X, that is, $x \wedge y = \inf_X\{x, y\}$ and $x \vee y = \sup_X\{x, y\}$. The empty subset is vacuously a sublattice of every lattice.

The definition implies immediately that the meet and join operations are commutative and associative, and for every $x, y \in X$, $x \wedge y \preceq x \vee y$. Moreover, for every $x, y \in X$, $x \preceq y \Leftrightarrow x \wedge y = x \Leftrightarrow x \vee y = y$. Furthermore, it follows immediately that a subset that is a sublattice is also a lattice (in the relative partial order), but the converse is not necessarily true, as shown below.

Examples 2.5 Consider the following examples.

1. Every subset of a chain is a sublattice of the chain.
2. Finite-dimensional Euclidean space, denoted $\mathbb{R}^N$, is a lattice in the product partial order. The subset of basis vectors in $\mathbb{R}^N$ is neither a sublattice nor a lattice (in the relative partial order).
3. The four point set $X = \{(0, 0), (1, 0), (0, 1), (1, 1)\} \subset \mathbb{R}^2$ is a sublattice of $\mathbb{R}^2$, and therefore, a lattice (in the relative partial order). The subsets $A = \{(0, 0), (1, 0)\}$ and $B = \{(0, 0), (0, 1)\}$ are sublattices (both in $\mathbb{R}^2$ and in X).
4. The four point set $X = \{(-1, 0), (1, 0), (0, 1), (2, 1)\} \subset \mathbb{R}^2$ is not a sublattice of $\mathbb{R}^2$, because $\inf_{\mathbb{R}^2}\{(1, 0), (0, 1)\} = (0, 0) \notin X$ (and also $\sup_{\mathbb{R}^2}\{(1, 0), (0, 1)\} = (1, 1) \notin X$). On the other hand, X is a lattice (in the relative partial order), because $\inf_X\{(1, 0), (0, 1)\} = (-1, 0) \in X$ and $\sup_X\{(1, 0), (0, 1)\} = (2, 1) \in X$.
5. The set of real-valued functions on X, $(\mathbb{R}^X, \preceq)$ is a lattice with meet and join defined pointwise, as above. If $X = \mathbb{R}$, then the set of continuous real-valued functions on X is a sublattice, but the set of differentiable functions is not a sublattice, because if $f(x) = x$ and $g(x) = -x$, then neither $f \wedge g$ nor $f \vee g$ are differentiable.
6. The power set $\mathcal{P}(X)$ of a set X is a lattice with meet and join operations given by set intersection and set union, as above.
7. If $(X, \preceq_X)$ and $(Y, \preceq_Y)$ are lattices, then the Cartesian product is a lattice (in the product partial order). For every (x, y) and (x', y') in $X \times Y$,

$$(x, y) \wedge (x', y') = (x \wedge x', y \wedge y') \text{ and } (x, y) \vee (x', y') = (x \vee x', y \vee y').$$

In other words, the Cartesian product of lattices is a lattice with meet and join operations defined componentwise.

8. Every chain (in the relative partial order) in a lattice X is a sublattice of X.

Proposition 2.6 *If A and B are sublattices of a lattice X, then $A \cap B$ is a sublattice of X.*

Proof If $A \cap B = \emptyset$ or a singleton, then it is trivially a sublattice. Otherwise, consider $x, y \in A \cap B$. Then $x \wedge y$ and $x \vee y$ are in A, because A is a sublattice, and $x \wedge y$ and $x \vee y$ are in B, because B is a sublattice, and therefore, $x \wedge y$ and $x \vee y$ are in $A \cap B$. □

A similar argument shows that intersection of arbitrarily many sublattices is a sublattice. The union of two sublattices is not necessarily a sublattice or a lattice (in the relative partial order). For example, consider $X = \{(0,0), (1,0), (0,1), (1,1)\}$. Then $A = \{(0,0), (0,1)\}$ and $B = \{(0,0), (1,0)\}$ are sublattices, but $A \cup B$ is neither a sublattice nor a lattice in the relative partial order.

2.1.4 *Interval*

For elements x, y in poset $(X, \preceq)$, let $[x, \infty) = \{\xi \in X \mid x \preceq \xi\}$, $(-\infty, x] = \{\xi \in X \mid \xi \preceq x\}$, and $[x, y] = \{\xi \in X \mid x \preceq \xi \preceq y\}$. An ***interval*** in X is a set of the form $(-\infty, x]$, or $[x, \infty)$, or $[x, y]$.

Examples 2.7 Consider the following examples.

1. In $\mathbb{R}$, the closed intervals $[0, 1]$, or $[0, \infty)$, or $(-\infty, 0]$ are intervals in the definition here. The set $A = \{1, 2, \ldots, 100\}$ is not an interval in $\mathbb{R}$, but is the interval $[1, 100]$ in the poset $\mathbb{N}$.
2. In $\mathbb{R}^2$, the interval $[(0,0), (1,1)]$ is the unit square. The diagonal in the unit square, $A = \{(t,t) | 0 \leq t \leq 1\}$, is not an interval, because $(0,0) \preceq (0,1) \preceq (1,1)$, but $(0,1) \notin A$. However, A is a sublattice and a chain, both as a subset of $\mathbb{R}^2$ and as a subset of $[(0,0), (1,1)]$. On the other hand, the unit line segment on the x-axis given by $[(0,0), (1,0)]$ is an interval in $\mathbb{R}^2$.
3. In the lattice of real-valued functions on X, $(\mathbb{R}^X, \preceq)$, for every $f, g \in X$, the interval $[f, \infty)$ consists of all functions that lie above

f pointwise, and the interval $[f, g]$ consists of all functions that lie between f and g pointwise.

4. Consider the set of distributions on $\mathbb{R}$, denoted X. That is, $X =$

$$\{F : \mathbb{R} \to [0, 1] \mid F \text{ is (weakly) increasing,} \\ \lim_{x \to -\infty} F(x) = 0, \lim_{x \to \infty} F(x) = 1\}.$$

   This is a lattice with pointwise partial order. For a distribution $F \in X$, every distribution in the interval $(-\infty, F]$ first order stochastically dominates F, every distribution in the interval $[F, \infty)$ is first order stochastically dominated by F, and every distribution in $[F, G]$ first order stochastically dominates G and is first order stochastically dominated by F.

5. In the power set of X, $\mathcal{P}(X)$, for $A \subset X$, the interval $[A, \infty)$ is the collection of all supersets of A, the interval $(-\infty, A]$ are all subsets of A, and the interval $[A, B]$ is the collection of all sets that are supersets of A and subsets of B.

Proposition 2.8 *Let X be a lattice.*

1. *For every* $x \in X$, $(-\infty, x] \neq \emptyset$ *and* $[x, \infty) \neq \emptyset$.
2. *For every* $x, y \in X$, $[x, y] \neq \emptyset$, *if, and only if,* $x \preceq y$.
3. *If* $A = [\hat{a}, \tilde{a}]$ *and* $B = [\hat{b}, \tilde{b}]$, *then* $A \cap B = [\hat{a} \vee \hat{b}, \tilde{a} \wedge \tilde{b}]$. *In particular,* $A \cap B$ *is nonempty, if, and only if,* $\hat{a} \vee \hat{b} \preceq \tilde{a} \wedge \tilde{b}$.

Proof Statement (1) follows immediately from the definition. In statement (2), sufficiency follows because if $z \in [x, y]$, then $x \preceq z \preceq y$, and necessity follows immediately from the definition. Statement (3) follows from $\hat{a} \preceq x \preceq \tilde{a}$ and $\hat{b} \preceq x \preceq \tilde{b}$, if, and only if, $\hat{a} \vee \hat{b} \preceq x \preceq \tilde{a} \wedge \tilde{b}$. □

For a subset A of lattice X, suppose $\inf_X A$ and $\sup_X A$ exist in X. The ***interval determined by*** A, (or ***interval generated by*** A,) denoted $[A]$, is defined as $[A] = [\inf_X A, \sup_X A]$. Statement (3) in Proposition 2.8 (page 19) can be used to show that $[A]$ is the intersection of all intervals in X that contain A, and is therefore, the smallest interval in X that contains A. The interval determined by the empty set is defined to be the empty set.

As shown in the following chapters, understanding the relation of products of generated intervals to the interval generated by the product turns out to be a new and useful tool in the study of monotone games. The following new result gives conditions under which these two are the same.

Proposition 2.9 *Let X be a lattice and A a subset of X with* $\inf A \in A$ *and* $\sup A \in A$*, and let Y be a lattice and B a subset of Y with* $\inf B \in B$ *and* $\sup B \in B$. *Then* $[A \times B] = [A] \times [B]$, *with the product partial order.*

Proof If either A or B is empty, the statement is trivially true. Otherwise, in one direction, $A \subset [A]$ and $B \subset [B]$ imply $A \times B \subset [A] \times [B]$, and as $[A] \times [B]$ is the interval given by $[(\inf A, \inf B), (\sup A, \sup B)]$, it follows that $[A \times B] \subset [A] \times [B]$. In the other direction, if $(a, b) \in [A] \times [B]$, then $\inf A \preceq a \preceq \sup A$ and $\inf B \preceq b \preceq \sup B$, and therefore, $(\inf A, \inf B) \preceq (a, b) \preceq (\sup A, \sup B)$ in the product partial order. As $(\inf A, \inf B) \in A \times B$ and $(\sup A, \sup B) \in A \times B$, it follows that $(a, b) \in [(\inf A, \inf B), (\sup A, \sup B)] \subset [A \times B]$. □

Proposition 2.10 *Every interval in a lattice X is a sublattice of X.*

Proof Consider a lattice X and an interval in X of the form $[x, y]$. Let $a, b \in [x, y]$. Then $x \preceq a$ and $x \preceq b$ implies $x \preceq a \wedge b$, and $a \preceq y$ and $b \preceq y$ implies $a \vee b \preceq y$. Combined with $a \wedge b \preceq a \vee b$, it follows that $a \wedge b$ and $a \vee b$ are in $[x, y]$. The cases $(-\infty, x]$ and $[x, \infty)$ are proved similarly. □

2.1.5 *Complete Lattice*

A lattice $(X, \preceq)$ is ***complete***, if for every nonempty subset A of X, $\inf_X A \in X$ and $\sup_X A \in X$. It follows that if X is a complete lattice, then $\inf_X X \in X$ and $\sup_X X \in X$, and therefore, $X = [\inf_X X, \sup_X X]$.

Subset A of a lattice X is ***subcomplete in*** X, if for every nonempty subset B of A, $\inf_X B \in A$ and $\sup_X B \in A$. Trivially, if X is complete, then X is subcomplete in itself. The empty set is vacuously complete and subcomplete in every lattice. If A is subcomplete in X, then A is a complete lattice in the relative partial order from X.

Examples 2.11 Consider the following examples.

1. Every finite lattice is complete, because a lattice is closed under finite iterations of $\wedge$ and $\vee$.

2. In $\mathbb{R}$, intervals of the form $[a, b]$ are subcomplete, whereas intervals of the form $(a, b]$, $[a, b)$, or (a, b) are not complete in the relative partial order. The subset $\{1, \ldots, 10\}$ is subcomplete.
3. The four point set $\{(0, 0), (1, 0), (0, 1), (1, 1)\} \subset \mathbb{R}^2$ is subcomplete in $\mathbb{R}^2$. The four point set $\{(-1, 0), (1, 0), (0, 1), (2, 2)\}$ is not subcomplete in $\mathbb{R}^2$, even though it is a complete lattice in the relative partial order. In other words, a subset of a lattice X may be a complete lattice in the relative partial order and not subcomplete in X.
4. In $\mathbb{R}^2$, the unit square given by the interval $[(0, 0), (1, 1)]$ is subcomplete. The diagonal in the unit square, $A = \{(t, t)|0 \leq t \leq 1\}$ is subcomplete as well (both in $\mathbb{R}^2$ and in $[(0, 0), (1, 1)]$), even though it is not an interval.
5. In the set of real-valued functions on X, $(\mathbb{R}^X, \preceq)$, the interval $[f, g]$ of all functions that lie between f and g pointwise is subcomplete.
6. In the power set of X, $\mathcal{P}(X)$, the interval $[A, B]$ of all supersets of A and subsets of B is subcomplete.
7. If X is a lattice, then every finite sublattice of X is subcomplete. This does not generalize to infinite sublattices even if they are chains in complete lattices. For example, let $X = [0, 1] \subset \mathbb{R}$ and $A = (0, 1]$. Then A is not subcomplete in X and not complete in the relative partial order.

Proposition 2.12 *If A and B are subcomplete in lattice X, then $A \cap B$ is subcomplete in X.*

Proof If $A \cap B = \emptyset$, then the statement is trivially true. Otherwise, consider a nonempty $C \subset A \cap B$. Then $C \subset A$ implies $\inf_X C \in A$ and $\sup_X C \in A$, and $C \subset B$ implies $\inf_X C \in B$ and $\sup_X C \in B$, and therefore, $\inf_X C \in A \cap B$ and $\sup_X C \in A \cap B$. □

An almost identical proof shows that intersection of arbitrarily many subcomplete lattices is subcomplete. The union of two subcomplete lattices is not necessarily even a sublattice, as shown by $\{(0, 0), (1, 0)\}$ and $\{(0, 0), (0, 1)\}$ in $\mathbb{R}^2$.

The four point set $\{(-1, 0), (1, 0), (0, 1), (2, 2)\}$ in $\mathbb{R}^2$ shows that it is not necessarily true that if A is a complete lattice in the relative partial order on X, then A is subcomplete in X.

The following result shows that intervals in a complete lattice are subcomplete.

Theorem 2.13 *Every interval in a complete lattice is subcomplete.*

Proof Consider a complete lattice $X = [\inf_X X, \sup_X X]$ and an interval $A = [a, b] \subset X$. If A is empty, it is vacuously subcomplete. Let $B \subset A$ be nonempty. As X is complete, $\inf_X B \in X$ and $\sup_X B \in X$. Moreover, $\forall x \in B, a \preceq x \preceq b$ implies that $a \preceq \inf_X B \preceq \sup_X B \preceq b$. Consequently, $\inf_X B \in [a, b]$ and $\sup_X B \in [a, b]$. □

In $\mathbb{R}^N$, there is the following useful characterization of subcompleteness in terms of compact sublattices due to Topkis (1998). The statement is rephrased for every subset and the proof is expanded to provide more detail. The following is helpful to prove this result. Recall that the standard topology on $\mathbb{R}^N$ is equivalent to the N-fold product topology on $\mathbb{R}$ and that compactness of a subset in $\mathbb{R}^N$ is equivalent to the subset being closed and bounded. Moreover, for a sequence (x^n) in $\mathbb{R}^N$, if we let $\liminf_n x^n = \bigvee_{n=1}^{\infty} \bigwedge_{k=n}^{\infty} x^k$ and $\limsup_n x^n = \bigwedge_{n=1}^{\infty} \bigvee_{k=n}^{\infty} x^k$, then x^n converges to x, if, and only if, $x = \liminf_n x^n = \limsup_n x^n$.

Theorem 2.14 *For every subset X of $\mathbb{R}^N$, X is subcomplete, if, and only if, X is a compact sublattice.*

Proof An empty set X satisfies the statement vacuously. For sufficiency, if X is subcomplete, then it is a sublattice. Moreover, $\inf_{\mathbb{R}^N} X \in X$ and $\sup_{\mathbb{R}^N} X \in X$ implies that X is bounded. To see that X is closed, consider a sequence $(x^n)_{n=1}^{\infty}$ in X such that $\lim_n x^n = x$. Then $x = \liminf_n x^n = \bigvee_{n=1}^{\infty} \bigwedge_{k=n}^{\infty} x^k$. Subcompleteness implies that for every n, $\bigwedge_{k=n}^{\infty} x^k \in X$ and using subcompleteness one more time, $x = \bigvee_{n=1}^{\infty} \bigwedge_{k=n}^{\infty} x^k \in X$. This shows that X is closed. As X is closed and bounded subset of $\mathbb{R}^N$, it is compact.

For necessity, suppose X is a compact sublattice and let A be a nonempty subset of X. Let B be the closure of A, that is, B is the union of A and the accumulation points of A. Then B is a closed subset of X hence compact. For each $i = 1, \ldots, N$, let $B_i = \{\xi \in \mathbb{R} \mid \exists x \in B, x_i = \xi\}$, let $\underline{b}_i = \inf B_i$, and notice that B is compact and nonempty implies $\underline{b}_i \in B_i$. Let $b^i \in B$ be such that its i-th component $b_i^i = \underline{b}_i$, and let $\underline{b} = b^1 \wedge b^2 \wedge \cdots \wedge b^n$. As X is a sublattice, it follows that $\underline{b} \in X$. By construction, $\underline{b}$ is a lower bound for B, and therefore, a lower bound for A. To see that $\underline{b} = \inf_{\mathbb{R}^N} A$, let z be a lower bound for A, that is, for every $a \in A$, $z \leq a$. As the set of points weakly higher than z is closed in $\mathbb{R}^N$ and contains A, it contains

B as well, and therefore, z is a lower bound for B. This implies that for every $b \in B$ and for every $i = 1, \ldots, N$, $z_i \leq b_i$, and therefore, for every $b \in B$ and for every $i = 1, \ldots, N$, $z_i \leq \underline{b}_i$, whence $z \leq \underline{b}$. It follows that $\inf_{\mathbb{R}^N} A = \underline{b} \in X$. Similarly, it can be shown that $\sup_{\mathbb{R}^N} A \in X$. □

2.1.6 Lattice Set Order

Let $(X, \preceq)$ be a lattice and A, B subsets of X. A ***is lower than*** B ***in the lattice set order***, denoted $A \sqsubseteq B$, if for every $a, b \in X$, $a \in A$ and $b \in B$ implies $a \wedge b \in A$ and $a \vee b \in B$. The lattice set order is sometimes termed the *Veinott set order*. It is attributed to Veinott (1989), as mentioned in Topkis (1978). Other terms used are the induced set ordering in Topkis (1998) and the strong set order in Milgrom and Shannon (1994). For convenience, B is higher than A in the lattice set order means the same as A is lower than B in the lattice set order.

Examples 2.15 Consider the following examples.

1. For singleton subsets $A = \{a\}$ and $B = \{b\}$ of lattice X, $A \sqsubseteq B \Leftrightarrow a \preceq b$.
2. In a lattice X consider the empty set $\emptyset$. It is vacuously true that for every subset A of X, $\emptyset \sqsubseteq A$ and $A \sqsubseteq \emptyset$.
3. In a lattice X, for every $x \preceq y$, $(-\infty, x] \sqsubseteq (-\infty, y]$ and $[x, \infty) \sqsubseteq [y, \infty)$. Moreover, for every $x \preceq x'$ and $y \preceq y'$, $[x, y] \sqsubseteq [x', y']$.
4. In $\mathbb{R}^2$, the square $[(0, 0), (5, 5)]$ is lower than the square given by $[(1, 1), (7, 7)]$ in the lattice set order, that is, $[(0, 0), (5, 5)] \sqsubseteq [(1, 1), (7, 7)]$. Similarly, if $A = \mathbb{N}^2$, then the discrete lattice given by $A \cap [(0, 0), (5, 5)]$ is lower than the discrete lattice given by $A \cap [(1, 1), (7, 7)]$ in the lattice set order.
5. In $\mathbb{R}^2$, consider an arbitrary nonempty subset A of the interval $[(0, 0), (1, 1)]$ and an arbitrary nonempty subset B of $[(2, 2), (3, 3)]$. Then $A \sqsubseteq B$.
6. In the set of real-valued functions on X, $\mathbb{R}^X$, the set of functions bounded below by 0 and above by 5 is lower than the set of functions bounded below by 1 and above by 7 in the lattice set order.
7. In the power set, $\mathcal{P}(X)$, consider $A \subset C \subset X$ and $B \subset D \subset X$. If $A \subset B$ and $C \subset D$, then $[A, C] \sqsubseteq [B, D]$.

The lattice set order has the following properties as proved in Topkis (1998).

Theorem 2.16 *For nonempty subsets A, B, C of lattice X,*

1. $A \sqsubseteq A \Leftrightarrow A$ *is a sublattice of* X
2. $A \sqsubseteq B$ *and* $B \sqsubseteq A \Rightarrow A = B$
3. $A \sqsubseteq B$ *and* $B \sqsubseteq C \Rightarrow A \sqsubseteq C$

Proof Statement 1 is easy to prove. For statement 2, suppose $A \sqsubseteq B$ and $B \sqsubseteq A$. Fix $x \in A$ and $y \in B$. Then $A \sqsubseteq B \Rightarrow x \vee y \in B$, and combined with $B \sqsubseteq A$, it follows that $x = (x \vee y) \wedge x \in B$. As x is arbitrary, $A \subset B$. Similarly, $B \subset A$.

For statement 3, suppose $A \sqsubseteq B$ and $B \sqsubseteq C$. Fix $a \in A$ and $c \in C$ arbitrarily. Pick $b \in B$ arbitrarily. Then $B \sqsubseteq C \Rightarrow b \wedge c \in B$, and $A \sqsubseteq B \Rightarrow a \vee (b \wedge c) \in B$, and $B \sqsubseteq C \Rightarrow [a \vee (b \wedge c)] \vee c \in C$. But $[a \vee (b \wedge c)] \vee c = a \vee [(b \wedge c) \vee c] = a \vee c$. Thus $a \vee c \in C$. Similarly, $A \sqsubseteq B \Rightarrow a \vee b \in B$, and $B \sqsubseteq C \Rightarrow (a \vee b) \wedge c \in B$, and $A \sqsubseteq B \Rightarrow a \wedge [(a \vee b) \wedge c] \in A$. But $a \wedge [(a \vee b) \wedge c] = [a \wedge (a \vee b)] \wedge c = a \wedge c$. Thus $a \wedge c \in A$. □

In other words, the lattice set order is a partial order on the collection of nonempty sublattices of a lattice.

Theorem 2.17 *Let A and B be nonempty subsets of lattice X.*

1. *If* $\sup_X A \preceq \inf_X B$, *then* $A \sqsubseteq B$ *(whenever* $\sup_X A$, $\inf_X B$ *exist)*
2. *If* $A \sqsubseteq B$, *then* $\inf_X A \preceq \inf_X B$ *and* $\sup_X A \preceq \sup_X B$ *(whenever* $\inf_X A$, $\inf_X B$ *exist, and whenever* $\sup_X A$, $\sup_X B$ *exist)*

Proof For statement 1, suppose $\sup_X A \preceq \inf_X B$. Then for every $a \in A$ and $b \in B$, $a \preceq b$, and therefore, $a \wedge b = a \in A$ and $a \vee b = b \in B$.

For statement 2, suppose $A \sqsubseteq B$ and $\inf_X A$, $\inf_X B$ exist. Fix $a \in A, b \in B$ arbitrarily. Then $a \wedge b \in A$ implies $\inf A \preceq a \wedge b \preceq b$. As b is an arbitrary element of B, this shows that $\inf A$ is a lower bound for B, and therefore, $\inf A \preceq \inf B$. Similarly, $a \vee b \in B$ implies $a \preceq a \vee b \preceq \sup B$. As a is an arbitrary element of A, this shows that $\sup B$ is an upper bound for A, and therefore, $\sup A \preceq \sup B$. □

In both statements in Theorem 2.17 (page 24), A and B are arbitrary nonempty subsets of X and it is not assumed that $\inf A \in A$, or $\sup A \in A$, or $\inf B \in B$, or $\sup B \in B$.

2.1.7 Correspondence and Fixed Point

Let X and Y be sets. A ***correspondence from*** X ***to*** Y, denoted $\Phi : X \rightrightarrows Y$, is a function from X to the power set of Y, $\Phi : X \to \mathcal{P}(Y)$. It is nonempty valued, if for every $x \in X$, $\Phi(x) \neq \emptyset$. It is singleton valued, if for every x in X, $\Phi(x)$ is a singleton subset of Y. A function is viewed as a correspondence that is singleton valued.

For a correspondence $\Phi : X \rightrightarrows X$, a point $x \in X$ is ***a fixed point of*** Φ, if $x \in \Phi(x)$. The ***fixed point set*** of Φ is $\mathcal{E}(\Phi) = \{x \in X \mid x \in \Phi(x)\}$. The notation may be abbreviated to $\mathcal{E}$ when convenient.

Suppose $(X, \preceq_X)$ is a poset and $(Y, \preceq_Y)$ is a lattice. A correspondence $\Phi : X \rightrightarrows Y$ is subcomplete valued, if for every $x \in X$, $\Phi(x)$ is subcomplete in Y. A correspondence $\Phi : X \rightrightarrows Y$ is ***(weakly) increasing***, if for every $x, x' \in X, x \preceq x' \Rightarrow \Phi(x) \sqsubseteq \Phi(x')$.

Let $(X, \preceq_X)$ and $(Y, \preceq_Y)$ be posets. A function $f : X \to Y$ is ***(weakly) increasing***, if for every, $x, x' \in X, x \preceq_X x' \Rightarrow f(x) \preceq_Y f(x')$.

As shown by Zhou (1994), under weak conditions, the fixed point set of a (weakly) increasing correspondence is a nonempty, complete lattice. The proof of Zhou (1994) given in Topkis (1998) is expanded to include more detail to make it more accessible.

Theorem 2.18 *Let* $(X, \preceq)$ *be a complete lattice and* Φ *a correspondence on* X.
If Φ *is (weakly) increasing, nonempty valued, and subcomplete valued, then*

1. Φ *has a fixed point.*
2. *The largest fixed point of* Φ *is given by*

$$\bar{x}^* = sup_X \left\{x \in X | \Phi(x) \cap \left[x, \sup X\right] \neq \emptyset\right\}.$$

3. *The smallest fixed point of* Φ *is given by*

$$\underline{x}^* = inf_X \{x \in X | \Phi(x) \cap [\inf X, x] \neq \emptyset\}.$$

4. $\mathcal{E}(\Phi)$ *is a nonempty, complete lattice (in the relative partial order).*

Proof For notational convenience, let $\underline{x} = \inf X$ and $\bar{x} = \sup X$. Notice that either statement (2) or (3) implies statement (1).

To prove statement (2), let $A = \{x \in X | \Phi(x) \cap [x, \bar{x}] \neq \emptyset\}$. Then $\underline{x} \in A$, because $\Phi(\underline{x}) \cap [\underline{x}, \bar{x}] = \Phi(\underline{x}) \neq \emptyset$. Let $\bar{x}^* = \sup_X A$, and as Φ is subcomplete valued, let $\bar{y} = \sup_X \Phi(\bar{x}^*) \in \Phi(\bar{x}^*)$ and $\bar{z} = \sup_X \Phi(\bar{y}) \in \Phi(\bar{y})$.

Notice that $x \in A$ implies $\Phi(x) \cap [x, \bar{x}] \neq \emptyset$ and therefore, $x \preceq \sup_X \Phi(x)$. Moreover, $x \in A$ implies $x \preceq \bar{x}^*$, and as Φ is (weakly) increasing, it follows that $\Phi(x) \sqsubseteq \Phi(\bar{x}^*)$, and consequently, $\sup_X \Phi(x) \preceq \bar{y}$. In other words, for every $x \in A$, $x \preceq \bar{y}$, and therefore, $\bar{x}^* \preceq \bar{y}$. Applying Φ is (weakly) increasing to $\bar{x}^* \preceq \bar{y}$ implies $\Phi(\bar{x}^*) \sqsubseteq \Phi(\bar{y})$, and therefore, $\bar{y} = \sup_X \Phi(\bar{x}^*) \preceq \bar{z}$. This implies that $\bar{z} \in [\bar{y}, \bar{x}]$. As $\bar{z} \in \Phi(\bar{y})$ as well, it follows that $\bar{y} \in A$, and therefore, $\bar{y} \preceq \bar{x}^*$. Combined with $\bar{x}^* \preceq \bar{y}$, it follows that $\bar{x}^* = \bar{y} \in \Phi(\bar{x}^*)$ is a fixed point of Φ.

To see that $\bar{x}^*$ is the largest fixed point of Φ, let x^* be an arbitrary fixed point of Φ. Then $x^* \in \Phi(x^*)$ and $x^* \in [x^*, \bar{x}]$ implies $\Phi(x^*) \cap [x^*, \bar{x}] \neq \emptyset$, whence $x^* \in A$, and therefore, $x^* \preceq \bar{x}^*$.

Statement (3) is proved similarly.

To prove statement (4), we already know that $\mathcal{E}(\Phi)$ is not empty. Let E be a nonempty subset of $\mathcal{E}(\Phi)$.

Let $\bar{e} = \sup_X E$, which exists, because X is complete. For each $e \in E$, $e \in \Phi(e)$ implies $e \preceq \sup_X \Phi(e)$, and moreover, $e \preceq \bar{e}$ implies $\Phi(e) \sqsubseteq \Phi(\bar{e})$, whence $\sup_X \Phi(e) \preceq \sup_X \Phi(\bar{e})$. This shows that $\sup_X \Phi(\bar{e})$ is an upper bound for E, and therefore, $\bar{e} \preceq \sup_X \Phi(\bar{e})$.

Consider the correspondence $\psi : [\bar{e}, \bar{x}] \rightrightarrows [\bar{e}, \bar{x}]$ given by $\psi(x) = \Phi(x) \cap [\bar{e}, \bar{x}]$. This correspondence is nonempty valued, because for each $\bar{e} \preceq x$, $\Phi(\bar{e}) \sqsubseteq \Phi(x)$, and therefore, $\bar{e} \preceq \sup_X \Phi(\bar{e}) \preceq \sup_X \Phi(x)$, from which it follows that $\sup_X \Phi(x) \in \Phi(x) \cap [\bar{e}, \bar{x}] = \psi(x)$. Moreover, ψ is subcomplete valued, because each $\psi(x)$ is an intersection of two subcomplete sets, and ψ is (weakly) increasing, because Φ is (weakly) increasing. Therefore, ψ satisfies the hypotheses of this theorem and we may apply statements (1), (2), and (3) to ψ.

In particular, $e^* = \inf_{[\bar{e}, \bar{x}]} \{x \in [\bar{e}, \bar{x}] | \psi(x) \cap [x, \bar{x}] \neq \emptyset\}$ is the smallest fixed point of ψ. As $e^* \in \psi(e^*) = \Phi(e^*) \cap [\bar{e}, \bar{x}]$, it follows that e^* is a fixed point of Φ, that is, $e^* \in \mathcal{E}(\Phi)$. Combined with $\bar{e} \preceq e^*$, it follows that e^* is an upper bound for E.

To see that $e^* = \sup_{\mathcal{E}(\Phi)} E$, consider an arbitrary fixed point $x^* \in \mathcal{E}(\Phi)$ such that x^* is an upper bound for E. Then $\bar{e} \preceq x^*$ implies $x^* \in [\bar{e}, \bar{x}]$, and x^* is a fixed point implies $x^* \in \Phi(x^*)$, and therefore, x^* is a fixed point of

ψ, whence $e^* \preceq x^*$. Consequently, $\sup_{\mathcal{E}(\Phi)} E = e^* \in \mathcal{E}(\Phi)$. Similarly, it can be shown that $\inf_{\mathcal{E}(\Phi)} E \in \mathcal{E}(\Phi)$. □

The theorem for (weakly) increasing functions due to Tarski (1955) follows as a special case.

Theorem 2.19 *Let $(X, \preceq)$ be a complete lattice.*
If $f : X \to X$ is a (weakly) increasing function, then

1. *f has a fixed point.*
2. *The largest fixed point of f is given by* $\sup_X \{x \in X | x \preceq f(x)\}$.
3. *The smallest fixed point of f is given by* $\inf_X \{x \in X | f(x) \preceq x\}$.
4. *$\mathcal{E}(f)$ is a nonempty, complete lattice (in the relative partial order).*

Proof The function f may be viewed as a singleton valued correspondence $\Phi : X \rightrightarrows X$ given by $\Phi(x) = \{f(x)\}$. This correspondence is (weakly) increasing, nonempty valued, and subcomplete valued, and the result follows from the previous theorem. □

The set of fixed points of a weakly increasing correspondence may not necessarily be subcomplete, even when the correspondence is singleton valued.

Example 2.20 Let

$$X = \{(0, 0), (1, 0), (2, 0), (0, 1), (1, 1), (2, 1)\} \subset \mathbb{R}^2,$$

and $f : X \to X$ be given by

$$f(0, 0) = (0, 0), f(1, 0) = (0, 0), f(2, 0) = (2, 0),$$
$$f(0, 1) = (0, 0), f(1, 1) = (1, 1), f(2, 1) = (2, 1).$$

Then f is (weakly) increasing and its set of fixed points is

$$\mathcal{E}(f) = \{(0, 0), (2, 0), (1, 1), (2, 1)\}.$$

$\mathcal{E}(f)$ is a complete lattice that is not subcomplete in X.

Let $(X, \preceq_X)$ and $(Y, \preceq_Y)$ be posets. A correspondence $\Phi : X \rightrightarrows Y$ is ***never increasing***, if for every $\hat{x}, \tilde{x} \in X$, if $\hat{x} \prec_X \tilde{x}$, then for every $\hat{y} \in \Phi(\hat{x})$ and for every $\tilde{y} \in \Phi(\tilde{x})$, $\hat{y} \not\prec_Y \tilde{y}$. If $f : X \to Y$ is a function, the definition specializes to that of a not increasing function, that is, $x \prec y \Rightarrow f(x) \nprec f(y)$. If $f : X \to Y$ is a function and Y is a chain, the definition specializes to that of a (weakly) decreasing function, that is, $x \prec y \Rightarrow f(y) \preceq f(x)$.

In contrast to (weakly) increasing correspondences, the fixed point set of never increasing correspondences is totally unordered. Roy and Sabarwal (2008) show this for the case where X is a compact, convex subset of finite-dimensional Euclidean space. The result here is more general. It requires only that the domain (and target) is a poset. There is no additional restriction for it to be finite-dimensional, infinite-dimensional, compact, or convex. It is not required to be a vector space, a topological space, a measure space, or to have additional mathematical structure beyond that of a partial order. It is not required to be a lattice. An additional generalization here is that the definition of never increasing correspondence allows for weak inequality.

Theorem 2.21 *For every poset $(X, \preceq)$ and for every never increasing correspondence $\Phi : X \rightrightarrows X$, the fixed point set of Φ is totally unordered.*

Proof If the fixed point set is empty or a singleton, the statement is trivially true. For the general case, define the sets $A = \{x \in X \mid \exists x' \in \Phi(x), x \preceq x'\}$ and $B = \{x \in X \mid \exists x' \in \Phi(x), x' \preceq x\}$, and notice that every fixed point of Φ is in $A \cap B$.

Moreover, every fixed point x of Φ is a maximal element of A, that is, there is no $a \in A$ such that $x \prec a$, and this can be seen as follows. If there is $a \in A$ with $x \prec a$, then $a \in A$ implies there is $x' \in \Phi(a)$ such that $a \preceq x'$. In other words, for $x \prec a$, there is $x \in \Phi(x)$ and there is $x' \in \Phi(a)$ such that $x \prec x'$, a contradiction to Φ is never increasing. Similarly, every fixed point x of Φ is a minimal element of B.

If x and y are two fixed points of Φ and are comparable, say, $x \prec y$, then x cannot be a maximal element of A and y cannot be a minimal element of B, and in either case, there is a contradiction. □

As proof of the theorem indicates, the fixed point set of a never increasing correspondence may be empty.

Examples 2.22 Consider $X = \{a, b, c, d\}$, with $a \prec b \prec d$, $a \prec c \prec d$, and b and c are unordered. Let $f : X \to X$ be given by

$$f(a) = d,\ f(b) = c,\ f(c) = b,\ f(d) = a.$$

It is easy to check that f is a (strictly) decreasing function on X, and therefore, $x \mapsto \{f(x)\}$ is a singleton valued, never increasing correspondence. Nevertheless, f does not have a fixed point.

Corollary 2.23 *Let $(X, \preceq)$ be a poset.*

1. *The fixed point set of every not increasing function on X is totally unordered.*
2. *The fixed point set of every (weakly) decreasing function on X is totally unordered.*

Proof A function $f : X \to X$ may be viewed as a singleton valued correspondence $\Phi : X \rightrightarrows X$ given by $\Phi(x) = \{f(x)\}$. In either statement, the correspondence is never increasing, and the result follows from Theorem 2.21 (page 28). □

Corollary 2.24 *For every poset $(X, \preceq)$ and for every never increasing correspondence $\Phi : X \rightrightarrows X$, the following are equivalent.*

1. *$\mathcal{E}(\Phi)$ is a nonempty, complete lattice (in the relative partial order)*
2. *$\mathcal{E}(\Phi)$ is a nonempty lattice (in the relative partial order)*
3. *Φ has exactly one fixed point*

Proof The implications (1) implies (2), and (3) implies (1) are trivial. To see that (2) implies (3), suppose $\mathcal{E}(\Phi)$ is a nonempty lattice (in the relative partial order). Then Φ has at least one fixed point. If Φ has more than one fixed point, say, x and y are fixed points of Φ with $x \neq y$, then $\mathcal{E}(\Phi)$ is a lattice implies that the fixed point given by their join, say z, is weakly larger than both x and y and is different from at least one of them. Therefore, either $x \prec z$ or $y \prec z$, and in either case, the fixed point set is not totally unordered, a contradiction. □

Corollary 2.25 *If X is a chain and $\Phi : X \rightrightarrows X$ is a never increasing correspondence, then Φ has exactly zero or one fixed point.*

Proof If Φ has two or more fixed points, say, x and y with $x \neq y$, then as X is a chain, x and y are comparable, a contradiction to the fact that the set of fixed points is totally unordered. □

Let $(X, \preceq_X)$ and $(Y, \preceq_Y)$ be posets. A correspondence $\Phi : X \rightrightarrows Y$ is ***never decreasing***, if for every $\hat{x}, \tilde{x} \in X$, if $\hat{x} \prec_X \tilde{x}$, then for every $\hat{y} \in \Phi(\hat{x})$ and for every $\tilde{y} \in \Phi(\tilde{x})$, $\tilde{y} \not\prec_Y \hat{y}$. If Φ is singleton valued, then the definition specializes to that of a not decreasing function, that is, $x \prec y \Rightarrow \Phi(y) \not\prec \Phi(x)$. Moreover, if X is a chain and Φ is singleton valued, then the definition specializes to that of a (weakly) increasing function, that is, $x \prec y \Rightarrow \Phi(x) \preceq \Phi(y)$.

2.2 Lattice Game

2.2.1 *Definition*

A ***lattice game*** is defined by $\Gamma = (X_i, u_i)_{i=1}^I$, where

1. The game has finitely many ***players***, I, indexed by $i \in \{1, \ldots, I\}$.
2. For each i, the ***action space of player*** i is X_i. It is assumed to be a nonempty, subcomplete lattice in $\mathbb{R}^{n_i}$ with the Euclidean partial order. The space of profiles of actions is $X = X_1 \times \cdots \times X_I$ with the product partial order. A typical element is denoted $x = (x_i)_{i=1}^I$. In order to distinguish between actions of player i and actions of opponents of i, X may be denoted as $X = X_i \times X_{-i}$, where $X_{-i} = \times_{j \neq i} X_j$, and typical element as $x = (x_i, x_{-i}) \in X_i \times X_{-i}$.
3. For each i, the ***payoff for player*** i is $u_i : X_i \times X_{-i} \to \mathbb{R}$. It depends on actions of player i, x_i, and those of opponents of i, x_{-i}, mapping (x_i, x_{-i}) to $u_i(x_i, x_{-i})$. For each i, u_i is upper semicontinuous on X and continuous on X_{-i}.

The definition of lattice game is a useful general construct that naturally subsumes monotone games. Several results are proved and viewed more clearly at this greater level of generality. The definition here is a special case of a general normal form game or strategic game. The main new specialization is that of partial order to formalize the idea of direction. When action space of each player is finite, semicontinuity and continuity conditions on the payoff function are satisfied automatically. In this case, the only additional requirement is the notion of order or direction formalized by a sublattice of finite-dimensional Euclidean space.

Although a player's action space in a lattice game is assumed to be finite dimensional, several results for lattice games generalize to Banach lattices or to Riesz spaces with natural additional assumptions. There are some variations among interval topology, order convergence topology, and metric space topology in more general spaces and this has a corresponding impact on continuity of payoff functions and compactness and completeness of a lattice. These complications do not arise in finite-dimensional action spaces, which remain a natural setting in many applications. This provides a sufficiently broad scope for the theory without invoking additional topological apparatus.

For more general results, the components of a lattice game are developed in their natural and more general settings in the previous section. Moreover, proofs of results for lattice games in this section are developed in a manner that shows their natural extension to more general settings with additional assumptions.

Each of the examples given in the introduction is a lattice game. In the first six examples, action space of each player has two actions, with one lower than the other. In the remaining two examples, each player has a (compact) interval of actions in the real numbers and payoff is jointly continuous in all variables.

Given a lattice game, it is useful to consider the same game but with actions restricted to a smaller set than in the original game. Consider a lattice game $\Gamma = (X_i, u_i)_{i=1}^I$. For every profile of actions $\hat{x} = (\hat{x}_1, \ldots, \hat{x}_I) \in X$ and $\hat{y} = (\hat{y}_1, \ldots, \hat{y}_I) \in X$ with $\hat{x} \preceq \hat{y}$, the ***lattice game restricted to*** $[\hat{x}, \hat{y}]$ is the lattice game $\hat{\Gamma} = (\hat{X}_i, \hat{u}_i)_{i=1}^I$, where for every i, $\hat{X}_i = [\hat{x}_i, \hat{y}_i]$ is a subset of X_i with the relative partial order from X_i, the space of profiles of actions is $\hat{X} = \hat{X}_1 \times \cdots \times \hat{X}_I$, and for every player i and for every $x \in \hat{X}$, $\hat{u}_i(x) = u_i(x)$. It follows immediately that $\hat{u}_i$ is upper semicontinuous on $\hat{X}$ and continuous on $\hat{X}_{-i}$ in the relative topology.

The ideas of best response, Nash equilibrium, serially undominated strategies, and dominance solvability are the standard ones.

2.2.2 Best Response

When opponents of player i play x_{-i}, the effect of player i action $x_i \in X_i$ on payoff of player i is quantified by $u_i(x_i, x_{-i})$. An action $x_i \in X_i$ is in the best interest of player i, if playing x_i gives player i the highest possible payoff given that opponents of i play x_{-i}. In other words, the ***best response set of player i to opponent actions*** x_{-i} is

$$B_i(x_{-i}) = \arg\max_{x_i \in X_i} u_i(x_i, x_{-i}).$$

Each element $x_i \in B_i(x_{-i})$ is a best response of player i to opponent actions x_{-i}. For every player i, as X_i is complete and u_i is upper semicontinuous on X, it follows that for every x_{-i}, $B_i(x_{-i})$ is nonempty and compact. The ***best response correspondence for player*** i is $B_i : X_{-i} \rightrightarrows X_i$, mapping x_{-i} to $B_i(x_{-i})$. The ***joint best response correspondence*** is $B : X \rightrightarrows X$, defined by

$$B(x) = B_1(x_{-1}) \times B_2(x_{-2}) \times \cdots \times B_I(x_{-I}),$$

the Cartesian product of each player's best response set. As each $B_i(x_{-i})$ is nonempty and compact, it follows that for every x, $B(x)$ is nonempty and compact.

As shown in the following chapters, some properties of monotone games depend on limiting behavior of different versions of best response dynamics. This limiting behavior depends both on monotonicity properties and on continuity properties. A new insight here is that these two properties can be separated to provide a unifying thread. A deeper analysis shows that upper hemicontinuity of best response correspondence is the relevant continuity property. Isolating and formalizing this property separately at the level of a lattice game provides unifying proofs across different classes of monotone games.

For subsets X, Y of potentially different Euclidean spaces, with Y compact, a correspondence $\Phi : X \rightrightarrows Y$ ***is upper hemicontinuous***, if for every sequence (x_n) in X and every sequence (y_n) in Y, if $x_n \to x$, $y_n \to y$, and $y_n \in \Phi(x_n)$, then $y \in \Phi(x)$. In other words, on compact target spaces, upper hemicontinuity is equivalent to the property that Φ has closed graph. More general definitions and characterizations of different continuity properties for correspondences are available in Klein (1984).

Upper semicontinuity and continuity properties on payoffs on compact action spaces imply that individual and joint best response correspondences are nonempty valued, compact valued, and upper hemicontinuous.

Theorem 2.26 *In every lattice game,*

1. *For every i, best response correspondence for player i is nonempty valued, compact valued, and upper hemicontinuous.*
2. *The joint best response correspondence is nonempty valued, compact valued, and upper hemicontinuous.*

Proof For statement (1), let $B_i : X_{-i} \rightrightarrows X_i$ be the best response correspondence for player i. As X_i is subcomplete , it follows from Theorem 2.14

(page 22) that X_i is compact, and as u_i is upper semicontinuous on X, it follows that for every x_{-i}, $B_i(x_{-i})$ is nonempty and compact. For upper hemicontinuity, let $(x^n_{-i})_{n=0}^{\infty}$ be a sequence in X_{-i}, $(y_i^n)_{n=0}^{\infty}$ be a sequence in X_i, $x^n_{-i} \to x_{-i}$, $y_i^n \to y_i$, and for every n, $y_i^n \in B_i(x^n_{-i})$. Suppose $y_i \notin B_i(x_{-i})$. Then there is action $z_i \in X_i$ such that $u_i(z_i, x_{-i}) > u_i(y_i, x_{-i})$. Let $\epsilon = u_i(z_i, x_{-i}) - u_i(y_i, x_{-i}) > 0$ and $\xi = u_i(y_i, x_{-i}) + \frac{\epsilon}{2}$. By upper semicontinuity of u_i on X, there is N_1 such that for all $n \geq N_1$, $u_i(y_i^n, x^n_{-i}) < \xi$. By continuity of u_i on X_{-i}, there is N_2 such that for all $n \geq N_2$, $\xi < u_i(z_i, x^n_{-i})$. Therefore, for all $n \geq \max(N_1, N_2)$, $u_i(y_i^n, x^n_{-i}) < u_i(z_i, x^n_{-i})$, which contradicts $y_i^n \in B_i(x^n_{-i})$.

For statement (2), let $B : X \rightrightarrows X$ be the joint best response correspondence. From statement (1), it follows that B is nonempty valued and compact valued. For upper hemicontinuity, let $(x^n)_{n=0}^{\infty}$, $(y^n)_{n=0}^{\infty}$ be sequences in X, $x^n \to x$, $y^n \to y$, and for every n, $y^n \in B(x^n)$. By statement (1), for every i, $y_i \in B_i(x_{-i})$, and therefore, $y \in B(x)$. □

Examples 2.27 Consider the following examples.

1. Consider the coordination game (Example 1.1, page 2). When player 2 plays A, the payoff to player 1 from playing A is $u_1(A, A) = 2$ and the payoff from playing B is $u_1(B, A) = 1$, and therefore, best response of player 1 is $B_1(A) = \{A\}$. Similarly, $B_1(B) = \{B\}$. This characterizes the best response correspondence for player 1, $B_1 : \{A, B\} \rightrightarrows \{A, B\}$. In this example, it is in the best interest of player 1 to coordinate or move in the same direction as player 2. The same is true for player 2. The joint best response correspondence is given by

$$B(A, A) = \{(A, A)\},\ B(A, B) = \{(B, A)\},$$
$$B(B, A) = \{(A, B)\},\ B(B, B) = \{(B, B)\}.$$

2. In dove hawk game (Example 1.2, page 3), best response of player 1 is given by $B_1(D) = \{H\}$ and $B_1(H) = \{D\}$, and similarly for player 2. It is in the best interest of player 1 to move in the direction opposite to player 2. The same is true for player 2. The joint best response correspondence is given by

$$B(D, D) = \{(H, H)\},\ B(D, H) = \{(D, H)\},$$
$$B(H, D) = \{(H, D)\},\ B(H, H) = \{(D, D)\}.$$

3. In matching pennies (Example 1.3, page 4), best response of player 1 is given by $B_1(H) = \{T\}$ and $B_1(T) = \{H\}$, and for player 2, $B_1(H) = \{H\}$ and $B_1(T) = \{T\}$. It is in the best interest of player 1 to move in the direction opposite to player 2, but it is in the best interest of player 2 to move in the same direction as player 1. The joint best response correspondence is given by

$$B(H, H) = \{(T, H)\},\ B(H, T) = \{(H, H)\},$$
$$B(T, H) = \{(T, T)\},\ \ B(T, T) = \{(H, T)\}.$$

4. In the team trust game (Example 1.4, page 5), best response of player 1 is given by $B_1(C) = \{D\}$ and $B_1(D) = \{D\}$, and it is the same for player 2. In this situation, the best choice for each player is to play D regardless of opponent's choice. The best response correspondence of each player is constant at $\{D\}$. The joint best response correspondence is given by

$$B(C, C) = \{(D, D)\},\ B(C, D) = \{(D, D)\},$$
$$B(D, C) = \{(D, D)\},\ B(D, D) = \{(D, D)\}.$$

5. In the multi-player coordination game (Example 1.5, page 7), for each player i, action 1 is a best response to actions of other players, if, and only if, $u_i(1, x_{-i}) \geq u_i(0, x_{-i})$, or equivalently, $\sum_{j=1, j\neq i}^{I} x_j \geq \frac{I-1}{2}$. In other words, for each player i, playing 1 is the best choice when at least half of the other players also play 1. Best response of player i is

$$B_i(x_{-i}) = \begin{cases} \{1\} & \text{if } \sum_{j\neq i} x_j > \frac{I-1}{2}, \\ \{0, 1\} & \text{if } \sum_{j\neq i} x_j = \frac{I-1}{2}, \\ \{0\} & \text{if } \sum_{j\neq i} x_j < \frac{I-1}{2}. \end{cases}$$

Best response of player i is (weakly) increasing in actions of other players, showing a more generalized situation in which each player has an incentive to move in the same direction as others.

6. In the multi-player externality game (Example 1.6, page 7), for each player i, action 1 is a best response to actions of other players, if, and only if, no other player plays 1. In other words,

$$B_i(x_{-i}) = \begin{cases} \{1\} & \text{if } \sum_{j\neq i} x_j, = 0, \\ \{0\} & \text{if } \sum_{j\neq i} x_j > 0. \end{cases}$$

Best response of player i is weakly decreasing in actions of other players, showing a more generalized situation in which each player has an incentive to move in the direction opposite to others.

7. In Bertrand oligopoly (Example 1.7, page 8), profit of firm i from setting price x_i when opponents set prices x_{-i} is given by $u_i(x_i, x_{-i}) = (x_i - c)(a - bx_i + \beta(\sum_{j \neq i} x_j))$. This is a differentiable and concave function in x_i, and maximizing with respect to x_i yields

$$B_i(x_{-i}) = \frac{1}{2b}\left(a - c + \beta\left(\sum_{j \neq i} x_j\right)\right).$$

Profit maximizing price for firm i is increasing in prices set by other firms (subject to maximum unit price possible given demand), showing a more generalized situation in which each firm has an incentive to move in the same direction as others.

8. In Cournot oligopoly (Example 1.8, page 9), profit of firm i from producing quantity x_i when opponents produce quantities x_{-i} is given by $u_i(x_i, x_{-i}) = \left(a - b(\sum_{i=1}^{I} x_i)\right) x_i - cx_i$. This is a differentiable and concave function in x_i, and maximizing with respect to x_i yields

$$B_i(x_{-i}) = \frac{1}{2b}\left(a - c - b\left(\sum_{j \neq i} x_j\right)\right).$$

Profit maximizing quantity for firm i is weakly decreasing in output of other firms (subject to nonnegative output constraint), showing a more generalized situation in which each firm has an incentive to move in the direction opposite to other firms.

2.2.3 Nash Equilibrium Set

A ***Nash equilibrium***, or ***pure strategy Nash equilibrium***, in the game Γ is a profile of actions $x^* = (x_i^*)_{i=1}^I$ such that for every player i, $x_i^* \in B_i(x_{-i}^*)$. It follows immediately that x^* is a Nash equilibrium, if, and only if, $x^* \in B(x^*)$. In other words, x^* is a Nash equilibrium, if, and only if, it is a fixed point of the joint best response correspondence.

The ***Nash equilibrium set*** of the game is the collection of all Nash equilibria of the game, or equivalently, all fixed points of the joint best response correspondence B. It is denoted $\mathcal{E} = \{x \in X \mid x \in B(x)\}$. A

lattice game with Nash equilibrium is a lattice game in which the Nash equilibrium set is nonempty. A ***lattice game with Nash equilibrium on subintervals*** is a lattice game in which for every nonempty interval $[x, y]$, the Nash equilibrium set of the lattice game restricted to $[x, y]$ is nonempty.

Many classes of games possess Nash equilibrium on subintervals.

A well-known class is that of lattice games with the additional features that action space of each player is convex and payoff function of each player is quasiconcave in own action. Existence of Nash equilibrium follows from Kakutani's theorem.

When action spaces are not necessarily convex or payoff functions are not necessarily quasiconcave in own action, other results may apply. Lattice games that fall in an appropriate class of aggregative games and its generalizations have Nash equilibrium on subintervals. This includes the well-known classes of congestion games in Rosenthal (1973), potential games in Monderer and Shapley (1996), pseudo-potential games in Dubey et al. (2006), and the general class of quasi-aggregative games in Jensen (2010). As shown in these papers, lattice games that belong to one or more of these classes have Nash equilibrium on subintervals.

On the other hand, as shown by matching pennies, not every lattice game has a Nash equilibrium.

It is useful to have a notion of Nash equilibrium in which the best response of every player is unique in equilibrium. In a lattice game $\Gamma = (X_i, u_i)_{i=1}^I$, a profile of actions $x^* = (x_i^*)_{i=1}^I$ is a ***strict Nash equilibrium***, if for every player i, $B_i(x_{-i}^*) = \{x_i^*\}$. The ***strict Nash equilibrium set*** is the collection of all strict Nash equilibria in the game. It follows immediately that every strict Nash equilibrium is a Nash equilibrium. The converse is not true in general, but if best response correspondence is singleton valued, then every Nash equilibrium is a strict Nash equilibrium.

In some games, there is a symmetry in the incentives of players and it is useful to know if there is an equilibrium in which every player plays the same action. In a lattice game $\Gamma = (X_i, u_i)_{i=1}^I$, a profile of actions $x^* = (x_i^*)_{i=1}^I$ is a ***symmetric Nash equilibrium***, if it is a Nash equilibrium and for all players i and j, $x_i^* = x_j^*$.

Examples 2.28 Consider the following examples.

1. In the coordination game (Example 1.1, page 2), fixed points of the joint best response correspondence are (A, A) and (B, B), and therefore, the Nash equilibrium set is $\mathcal{E} = \{(A, A), (B, B)\}$. The Nash

equilibrium set is subcomplete (it is a chain) as a subset of the lattice of profiles of actions. Moreover, each Nash equilibrium is strict and each Nash equilibrium is symmetric.

2. In the dove hawk game (Example 1.2, page 3), the Nash equilibrium set is $\mathcal{E} = \{(D, H), (H, D)\}$. As these profiles are not comparable in the product partial order, the Nash equilibrium set is totally unordered. Each Nash equilibrium is strict.
3. The matching pennies game (Example 1.3, page 4) has no Nash equilibrium, $\mathcal{E} = \emptyset$.
4. The team trust game (Example 1.4, page 5) has a unique Nash equilibrium, $\mathcal{E} = \{(D, D)\}$.
5. In the multi-player coordination game (Example 1.5, page 7), there are exactly two Nash equilibria,

$$\mathcal{E} = \{(0, \ldots, 0), (1, \ldots, 1)\}.$$

The Nash equilibrium set is subcomplete as a subset of the lattice of profiles of actions $\{0, 1\}^I$. As in the earlier coordination game, both equilibria are strict (even though the best response correspondence is not singleton valued) and both equilibria are symmetric.

6. The multi-player externality game (Example 1.6, page 7) has I Nash equilibria, given by the basis vectors in $\mathbb{R}^I$.

$$\mathcal{E} = \{(1, 0, \ldots, 0), (0, 1, \ldots, 0), \ldots, (0, 0, \ldots, 1)\} \subset \{0, 1\}^I.$$

Each Nash equilibrium is strict and the Nash equilibrium set is totally unordered.

7. In Bertrand oligopoly (Example 1.7, page 8), the Nash equilibrium is unique, strict, and symmetric, with equilibrium price of each firm i given by $x_i^* = \frac{a-c}{2b-(I-1)\beta}$. The Nash equilibrium set is trivially a complete lattice.
8. In Cournot oligopoly (Example 1.8, page 9), the Nash equilibrium is unique, strict, and symmetric, with equilibrium output of each firm i given by $x_i^* = \frac{a-c}{I+1}$. The Nash equilibrium set is trivially a complete lattice.

2.2.4 Dominance Solvability and Global Stability

Let $\hat{X} = \hat{X}_1 \times \cdots \times \hat{X}_I \subset X$ be a subset of profiles of actions. For each i, the set of ***undominated responses of player i to*** $\hat{X}_{-i}$ is

$$\mathcal{U}_i(\hat{X}_{-i}) = \{x_i \in X_i \mid \forall x_i' \in X_i, \exists \hat{x}_{-i} \in \hat{X}_{-i}, u_i(x_i, \hat{x}_{-i}) \geq u_i(x_i', \hat{x}_{-i})\}.$$

In other words, action $x_i \in \mathcal{U}_i(\hat{X}_{-i})$ is an undominated response for player i, if for every action x_i' for player i, there is some profile of opponent actions $\hat{x}_{-i} \in \hat{X}_{-i}$ that makes playing x_i (weakly) better than playing x_i'.

The negation of this statement formalizes the definition of strictly dominated action. That is, an action $x_i \in X_i$ is strictly dominated, if there is $\hat{x}_i \in X_i$ such that for every $x_{-i} \in \hat{X}_{-i}$, $u_i(\hat{x}_i, x_{-i}) > u_i(x_i, x_{-i})$. An underlying behavioral assumption is that a reasonable player will never play a strictly dominated action, because there is another action that gives the player a strictly higher payoff *regardless of what opponents do*. Therefore, reasonable outcomes in a game must lie in the set of undominated responses.

A best response to a profile of opponent actions in $\hat{X}_{-i}$ satisfies this condition, and therefore, for every $x_{-i} \in \hat{X}_{-i}$, if $x_i \in B_i(x_{-i})$, then $x_i \in \mathcal{U}_i(\hat{X}_{-i})$. When viewed as an operator, $\mathcal{U}_i$ is an increasing operator, that is, $A \subset B \subset X_{-i} \Rightarrow \mathcal{U}_i(A) \subset \mathcal{U}_i(B)$.

The set of ***profiles of undominated responses to*** $\hat{X}$ is

$$\mathcal{U}(\hat{X}) = \mathcal{U}_1(\hat{X}_{-1}) \times \cdots \times \mathcal{U}_I(\hat{X}_{-I}),$$

the Cartesian product of each player's undominated responses. In particular, a profile of best responses to a profile of actions in $\hat{X}$ satisfies this condition, and therefore, for every $\hat{x} \in \hat{X}$, if $\tilde{x} \in B(\hat{x})$, then $\tilde{x} \in \mathcal{U}(\hat{X})$. When viewed as an operator, $\mathcal{U}$ is an increasing operator as well, that is, $A \subset B \subset X \Rightarrow \mathcal{U}(A) \subset \mathcal{U}(B)$.

Serially undominated strategies are defined iteratively. For $n = 0$ and for every player i, let $\mathcal{U}_i^0 = X_i$, and let $\mathcal{U}^0 = X_1 \times \cdots \times X_I = X$. For $n \geq 1$ and for every player i, let $\mathcal{U}_i^n = \mathcal{U}_i(\mathcal{U}_{-i}^{n-1})$, and let $\mathcal{U}^n = \mathcal{U}_1^{n-1} \times \cdots \times \mathcal{U}_I^{n-1}$. Notice that $\mathcal{U}_i(X_{-i}) \subset X_i$ and $\mathcal{U}(X) \subset X$ imply that for every player i, $(\mathcal{U}_i^n)_{n=0}^{\infty}$ is a decreasing sequence of sets and $(\mathcal{U}^n)_{n=0}^{\infty}$ is a decreasing sequence of sets as well. The set of ***serially undominated strategies for player*** i is given by

$$\mathcal{U}_i^{\infty} = \cap_{n=0}^{\infty} \mathcal{U}_i^n,$$

and the set of ***profiles of serially undominated strategies*** is given by

$$\mathcal{U}^\infty = \cap_{n=0}^\infty \mathcal{U}^n.$$

It follows immediately that $\mathcal{U}^\infty = \mathcal{U}_1^\infty \times \cdots \times \mathcal{U}_I^\infty$. A game is ***dominance solvable***, if $\mathcal{U}^\infty$ is a singleton.

Serially undominated strategies implicitly assume thoughtful players, thinking about other thoughtful players, using infinite iterations of introspection about reasonable actions to arrive at possible solutions to the game. The uniqueness of such a solution, formulated as dominance solvability, is a robust prediction for this type of behavior.

Theorem 2.29 *Let* Γ *be a lattice game and* $x, y \in X$.

1. *If* $x \in B(y)$ *and* $y \in B(x)$, *then* x *and* y *are profiles of serially undominated strategies.*
2. *If* x *is a Nash equilibrium, then* x *is a profile of serially undominated strategies.*

Proof For statement (1), suppose $x \in B(y)$ and $y \in B(x)$. Using induction, it is trivially true that $x, y \in \mathcal{U}^0$. Suppose $x, y \in \mathcal{U}^n$. Then $x \in B(y)$ and $y \in B(x)$ imply that for every i, $x_i \in B_i(y_{-i})$ and for every i, $y_i \in B_i(x_{-i})$, and therefore, for every i, $x_i \in \mathcal{U}_i^{n+1}$ and $y_i \in \mathcal{U}_i^{n+1}$, whence $x, y \in \mathcal{U}^{n+1}$. Consequently, $x, y \in \mathcal{U}^\infty$. Statement (2) follows from statement (1), because if x is a Nash equilibrium, we may take $y = x$ in statement (1). □

The notion of adaptive dynamics follows Milgrom and Roberts (1990). Consider a sequence of profiles of actions $(x^n)_{n=0}^\infty$ in X. For every $k > \hat{k}$, let $P(\hat{k}, k) = \{x^n \mid \hat{k} \leq n < k\}$ be the profile of play along the sequence from index $\hat{k}$ to index k (not including k), and let $[P(n, k)]$ be the interval in X determined by $P(n, k)$, that is,

$$[P(n, k)] = [\inf P(n, k), \sup P(n, k)].$$

Moreover, let $\mathcal{U}([P(n, k)])$ be profiles of undominated responses to $[P(n, k)]$ and $[\mathcal{U}([P(n, k)])]$ the interval determined by $\mathcal{U}([P(n, k)])$, that is,

$$[\mathcal{U}([P(n, k)])] = [\inf \mathcal{U}([P(n, k)]), \sup \mathcal{U}([P(n, k)])].$$

A sequence $(x^n)_{n=0}^{\infty}$ in X is an ***adaptive dynamic***, if for every $n \geq 0$, there is $k_n > n$, such that for every $k \geq k_n$, $x^k \in [\mathcal{U}([P(n,k)])]$.

Intuitively, future play in an adaptive dynamic is eventually an undominated response to the interval determined by past play, or failing that, it is in the interval determined by undominated responses. In this sense, an adaptive dynamic allows for a flexible path of play with eventually reasonable behavior by every player.

For $x^0 \in X$, a ***best response dynamic from*** x^0 is a sequence $(x^n)_{n=0}^{\infty}$, where for each $n \geq 1$, $x^n \in B(x^{n-1})$. A best response dynamic is necessarily an adaptive dynamic.

Theorem 2.30 *In a lattice game, for every $x^0 \in X$, every best response dynamic from x^0 is an adaptive dynamic.*

Proof Fix $x^0 \in X$ and let $(x^n)_{n=0}^{\infty}$ be a best response dynamic from x^0. For $n \geq 0$, let $k_n = n + 1$. Then for every $k \geq k_n$, $x^{k-1} \in P(n,k)$, and combined with $x^k \in B(x^{k-1})$, it follows that $x^k \in \mathcal{U}(P(n,k))$. Therefore, $x^k \in [\mathcal{U}([P(n,k)])]$. □

A lattice game Γ is ***globally stable***, if there is $x^* \in X$ such that every adaptive dynamic converges to x^*.

Theorem 2.31 *If a lattice game is globally stable, then it has a unique Nash equilibrium and every adaptive dynamic converges to this Nash equilibrium.*

Proof Suppose a lattice game is globally stable and let $x^* \in X$ be such that every adaptive dynamic converges to x^*. Fix $x^0 \in X$ and consider any best response dynamic $(x^n)_{n=0}^{\infty}$ that starts from x^0. Then for every n, $x^{n+1} \in B(x^n)$, and combined with $x^n \to x^*$, upper hemicontinuity implies $x^* \in B(x^*)$. Therefore, x^* is a Nash equilibrium. Consider an arbitrary Nash equilibrium $\hat{x}$. Then the sequence $(x^n)_{n=0}^{\infty}$ constant at $\hat{x}$ for every n is a best response dynamic starting from $\hat{x}$ and therefore, $\hat{x} = x^*$. Thus, x^* is the unique Nash equilibrium in the game. By definition of global stability, every adaptive dynamic converges to this Nash equilibrium. □

Adaptive dynamic implicitly assumes myopic players, adapting dynamically over time to past play, and using convergence of such behavior to arrive at a possible solution to the game. When all adaptive behavior converges to the same limit, that limiting behavior is a robust prediction for outcome in the game.

2.2.5 *Parameterized Lattice Games*

A lattice game formalizes the notion of decentralized decision-making with interdependent effects and provides a framework to understand incentives for each player to move in particular directions based on actions of other players.

More generally, well-being of a decision-maker may depend both on actions of others and on the decision-making environment itself.

In a coordination game, if there is a subsidy to adopt a particular technology, then payoff from adoption goes up and it may be the case that the non-subsidized technology is no longer an equilibrium prediction.

If two firms competing as Cournot duopoly are producing an output with a positive societal impact, say, wifi hotspots, and a policy maker subsidizes cost of inputs, there is an incentive for both firms to produce more wifi hotspots, and this may lead to a lower market price.

In these instances, magnitude of subsidy is a parameter of the decision-making environment. Given one level of subsidy, there is a particular prediction for each player's behavior and a particular collective equilibrium prediction. Given a different level of subsidy, there may be a different prediction for each player's behavior and a different equilibrium prediction.

Parameters are included in the decision-making environment by introducing a partially ordered set of parameters. Partial order helps to formalize direction of increase or decrease for the parameter. A ***parameterized lattice game*** is a collection $\Gamma = ((X_i, u_i)_{i=1}^I, T)$, where

1. There are finitely many ***players***, I, indexed by $i \in \{1, \ldots, I\}$.
2. For each i, the ***action space of player*** i is X_i. It is assumed to be a nonempty, subcomplete lattice in $\mathbb{R}^{n_i}$ with the Euclidean partial order. The space of profiles of actions is $X = X_1 \times \cdots \times X_I$ with the product partial order. The ***parameter space*** is $T \subset \mathbb{R}^n$. It is a poset with the Euclidean partial order.
3. For each i, the ***payoff for player*** i is $u_i : X_i \times X_{-i} \times T \to \mathbb{R}$. It depends on actions of player i, x_i, actions of opponents of i, x_{-i}, and the parameter t, mapping (x_i, x_{-i}, t) to $u_i(x_i, x_{-i}, t)$. For each i, $u_i(\cdot, \cdot, t)$ is upper semicontinuous on X and continuous on X_{-i}, for every t.

The parameter is viewed as summarizing the decision-making environment for each player. The effect of decision-making environment on well-being of each decision-maker is formalized by the dependence of each player's payoff on the parameter.

A parameterized lattice game Γ may be viewed as a collection of lattice games $\Gamma(t)$, one for each decision-making environment $t \in T$. The ***game at*** t, $\Gamma(t)$, is the lattice game with the same players and action spaces as Γ, and for each i, the payoff for player i is the section of u_i determined by t.

For each $t \in T$, we may analyze best response at t and Nash equilibrium at t using the same ideas as earlier applied to $\Gamma(t)$. In other words, for each $t \in T$, the ***best response set of player*** i ***to opponent actions*** x_{-i} at t is

$$B_i(x_{-i}, t) = \arg\max_{x_i \in X_i} u_i(x_i, x_{-i}, t).$$

The ***joint best response set to*** x ***at*** t is

$$B(x, t) = B_1(x_{-1}, t) \times B_2(x_{-2}, t) \times \cdots \times B_I(x_{-I}, t).$$

The ***Nash equilibrium set at*** t is

$$\mathcal{E}(t) = \{x \in X \mid x \in B(x, t)\}.$$

Moreover, a parameterized lattice game provides a framework to inquire how best choices change with the parameter and how equilibrium predictions change with the parameter. This may be used to model effects of changes in decision-making environment on player actions and equilibrium predictions.

The ***best response correspondence for player*** i is $B_i : X_{-i} \times T \rightrightarrows X_i$, mapping (x_{-i}, t) to $B_i(x_{-i}, t)$. The ***joint best response correspondence*** is $B : X \times T \rightrightarrows X$, defined by

$$B(x, t) = B_1(x_{-1}, t) \times B_2(x_{-2}, t) \times \cdots \times B_I(x_{-I}, t).$$

The effect of a change in decision-making environment on choices of player i is captured by the change in $B_i(x_{-i}, t)$ with respect to t, and the effect of a change in decision-making environment on choices of all players simultaneously is captured by the change in $B(x, t)$ with respect to t.

Table 2.1 Parameterized coordination game

		*P*2	
		A	*B*
*P*1	*A*	2, 1	0, t
	B	t, 0	$1+t, 2+t$

The ***Nash equilibrium correspondence*** (or, ***equilibrium correspondence***) is $\mathcal{E} : T \rightrightarrows X$, mapping t to $\mathcal{E}(t)$. This captures the effect of a change in decision-making environment on equilibrium outcomes.

In many situations, it is of interest to determine when is it the case that if a parameter goes up, the equilibrium profile of actions goes up as well. This is the idea of monotone comparative statics of equilibrium outcomes.

Example 2.32 (Parameterized coordination game) In the coordination game (Example 1.1, page 2), suppose technology B is better (or higher) than technology A, denoted $A \prec B$. Suppose that a policy maker provides an adoption benefit of t payoff units to each player who adopts technology B. What should the adoption benefit be so that both players adopt the better technology for sure?

The payoff bimatrix with adoption benefit t is given in Table 2.1 (page 43). With no adoption benefit (that is, $t = 0$), the Nash equilibrium set is $\mathcal{E}(0) = \{(A, A), (B, B)\}$. For adoption benefit $t > 1$, it is in player 2's best interest to choose technology B regardless of what player 1 chooses, and therefore, the Nash equilibrium set reduces to $\mathcal{E}(t) = \{(B, B)\}$. In other words, even though for $1 < t < 2$, technology B is not the dominant choice for player 1, the coordination incentive moves the equilibrium to the better (or higher) technology B.

A ***parameterized lattice game with Nash equilibrium*** is a parameterized lattice game in which for every $t \in T$, $\mathcal{E}(t)$ is nonempty. A ***parameterized lattice game with Nash equilibrium on subintervals*** is a parameterized lattice game in which for every $t \in T$, $\Gamma(t)$ is a lattice game with Nash equilibrium on subintervals, that is, for every nonempty interval $[x, y]$, the Nash equilibrium set of $\Gamma(t)$ restricted to $[x, y]$ is nonempty.

This chapter presents some of the mathematical framework useful for the study of lattice games. This material is helpful throughout the book.

The following chapters present additional mathematical framework useful to study monotone incentives and monotone games. Chapter 3 (page 45) studies codirectional incentives and games with strategic complements. Chapter 4 (page 79) studies contradirectional incentives and games with strategic substitutes. Chapter 5 (page 131) studies games in which some players may have strategic complements and others may have strategic substitutes. Monotone games encapsulate these three classes of games.

References

Birkhoff, G. (1995). *Lattice Theory*. American Mathematical Society (3rd ed.).

Dubey, P., Haimanko, O., & Zapechelnyuk, A. (2006). Strategic Complements and Substitutes, and Potential Games. *Games and Economic Behavior*, *54*, 77–94.

Jensen, M. K. (2010). Aggregative Games and Best-Reply Potentials. *Economic Theory*, *43*(1), 45–66.

Klein, E. (1984). *Theory of Correspondences: Including Applications to Mathematical Economics*. Canadian Mathematical Society Series of Monographs and Advanced Texts.

Milgrom, P., & Roberts, J. (1990). Rationalizability, Learning, and Equilibrium in Games with Strategic Complementarities. *Econometrica*, *58*(6), 1255–1277.

Milgrom, P., & Shannon, C. (1994). Monotone Comparative Statics. *Econometrica*, *62*(1), 157–180.

Monderer, D., & Shapley, L. (1996). Potential Games. *Games and Economic Behavior*, *14*, 124–143.

Rosenthal, R. W. (1973). A Class of Games Possessing Pure-Strategy Nash Equilibria. *International Journal of Game Theory*, *2*(1), 65–67.

Roy, S., & Sabarwal, T. (2008). On the (Non-)Lattice Structure of the Equilibrium Set in Games with Strategic Substitutes. *Economic Theory*, *37*(1), 161–169.

Tarski, A. (1955). A Lattice-theoretical Fixpoint Theorem and Its Application. *Pacific Journal of Mathematics*, *5*(2), 285–309.

Topkis, D. (1978). Minimizing a Submodular Function on a Lattice. *Operations Research*, *26*, 305–321.

Topkis, D. (1998). *Supermodularity and Complementarity*. Princeton: Princeton University Press.

Veinott, A. F. (1989). Lattice Programming (Unpublished Notes from Lectures Delivered). Johns Hopkins University.

Zhou, L. (1994). The Set of Nash Equilibria of a Supermodular Game Is a Complete Lattice. *Games and Economic Behavior*, *7*(2), 295–300.

CHAPTER 3

Games with Strategic Complements

Abstract Codirectional incentives formalize situations in which participants have an incentive to move in the same direction as other participants. Payoff functions with a combination of increasing differences, supermodularity, single crossing property, and quasisupermodularity capture these incentives. Games with strategic complements are lattice games in which each player has an incentive to move in the same direction as their opponents. Special features of these games imply particular patterns for individual and equilibrium behavior. The theory identifies principles governing behavior in these cases and a diversity of examples highlight applications of these principles.

Keywords Codirectional incentives · Strategic complements · Increasing maximizers · Reinforcing direct and indirect effects · Complete lattice equilibrium set · Increasing extremal equilibria

As discussed in Chapter 1 (page 1), ideas of coordination and opposition arise in many socioeconomic situations.

The role of strategic complements and strategic substitutes in economics was highlighted in contemporaneous papers by Fudenberg and Tirole (1984) and Bulow et al. (1985). They studied these incentives when there are two firms in an industry. A standard insight in previous work was that firms may overinvest strategically to deter entry by competitors, as shown

T. Sabarwal, *Monotone Games*,
https://doi.org/10.1007/978-3-030-45513-2_3

in Spence (1977), Dixit (1979), Dixit (1980), and others. In contrast, Fudenberg and Tirole (1984) and Bulow et al. (1985) show that this result depends on whether there are strategic complements or strategic substitutes. In particular, with strategic complements, firms may end up overinvesting to deter entry, and with strategic substitutes, firms may end up underinvesting so they can compete more aggressively with new entrants. These papers provide additional examples of the effect of strategic complements and strategic substitutes on competitive strategy.

The economic intuition behind these applications kindled broader interest in the mathematical principles underlying strategic complements and strategic substitutes.

For about a quarter century since then, the main focus was on games with strategic complements (GSC). Early work includes that of Topkis (1978) and Topkis (1979). Results for several general cases are available in Lippman et al. (1987), Sobel (1988), Milgrom and Roberts (1990), Vives (1990), Zhou (1994), Echenique (2002), Echenique and Sabarwal (2003), and Amir et al. (2008), among others. Book length treatments are Topkis (1998) and Vives (1999).

This chapter presents foundations of the theory of games with strategic complements as developed in this literature.

The general theory of codirectional incentives is developed with different characterizations and connections among the various components and with many illustrative examples. This is used to present conditions under which best response of a player is increasing in opponent actions, and this helps, in turn, to define strategic complements.

Games with strategic complements have several appealing features. In every GSC, the Nash equilibrium set in a nonempty, complete lattice. This provides an equilibrium existence theorem without relying on convexity and continuity properties. It also provides a complete lattice structure to the Nash equilibrium set, which guarantees the existence of a smallest Nash equilibrium and a largest Nash equilibrium.

The best response dynamic starting from inf X provides a computational algorithm to compute the smallest Nash equilibrium and the best response dynamics starting from sup X provides a computational algorithm to compute the largest Nash equilibrium. These Nash equilibria bound the set of serially undominated strategies and limiting behavior of adaptive dynamics. Therefore, uniqueness of Nash equilibrium is equivalent to dominance solvability and global stability.

In parameterized GSC, when environmental parameters provide players an incentive to take higher action, strategic complements reinforce this incentive. Consequently, the action of every player in the smallest equilibrium goes up with the parameter and the action of every player in the largest equilibrium goes up with the parameter. This is the well-known result for monotone comparative statics of extremal equilibrium outcomes. This result does not imply that every equilibrium at a higher parameter is higher than every equilibrium at the lower parameter, even for symmetric Nash equilibrium.

3.1 Codirectional Incentives

Strategic complements formalize a class of interdependent decisions that exhibit patterns of coordination, or movement in the same direction as other decision-makers. As mentioned in Chapter 1 (page 1), this basic principle manifests in many seemingly unrelated socioeconomic outcomes such as bank runs, speculative currency attacks, political uprisings, social network success or failure, run on groceries in a pandemic, online shopping platform success or failure, segregation and desegregation, propagation of health epidemics or misinformation on networks, coordinating successful social events, and so on.

One way to think about incentive for a player to take a higher action when opponents take a higher action is that marginal benefit to the player from a higher action is greater when opponents take a higher action. This intuition is formalized by the property of increasing differences. The notion of increasing differences generalizes to that of supermodular function on a lattice.

Increasing differences and supermodularity are both cardinal properties that do not necessarily hold under monotonic transformations of a payoff function. The more general notions of single crossing property and quasisupermodular function provide this additional flexibility. These properties provide a mathematical foundation for codirectional incentives.

Single crossing property and quasisupermodularity are used to characterize when maximizers in an optimization problem are increasing in the parameters of the problem. They provide a mathematical foundation for strategic complements and for statements of the form: If actions of opponents go up, then the best choice of a given player goes up as well.

3.1.1 *Increasing Differences*

For posets $(X, \preceq_X)$ and $(Y, \preceq_Y)$, a function $f : X \times Y \to \mathbb{R}$ has ***increasing differences in*** (x, y), if for every $\hat{x} \preceq_X \tilde{x}$ and for every $\hat{y} \preceq_Y \tilde{y}$, $f(\tilde{x}, \hat{y}) - f(\hat{x}, \hat{y}) \leq f(\tilde{x}, \tilde{y}) - f(\hat{x}, \tilde{y})$. In other words, for every $\hat{x} \preceq_X \tilde{x}$, the difference $f(\tilde{x}, y) - f(\hat{x}, y)$ is (weakly) increasing in y.

In the setting of a lattice game, $(X_i, u_i)_{i=1}^I$, if player i payoff has increasing differences in (x_i, x_{-i}), the marginal benefit to player i from taking a higher action over a lower action, $u_i(\tilde{x}_i, x_{-i}) - u_i(\hat{x}_i, x_{-i})$, is an increasing function of opponent actions. This provides player i an incentive to take a higher action when opponents take higher actions.

Similarly, a function $f : X \times Y \to \mathbb{R}$ has *increasing differences in* (y, x), if for every $\hat{x} \preceq_X \tilde{x}$ and for every $\hat{y} \preceq_Y \tilde{y}$, $f(\hat{x}, \tilde{y}) - f(\hat{x}, \hat{y}) \leq f(\tilde{x}, \tilde{y}) - f(\tilde{x}, \hat{y})$. In other words, for every $\hat{y} \preceq_Y \tilde{y}$, $f(x, \tilde{y}) - f(x, \hat{y})$ is (weakly) increasing in x. Rearranging terms shows that f has increasing differences in (x, y), if, and only if, f has increasing differences in (y, x). The order of x and y in the definition does not matter.

In lattice games, payoff functions are on products of lattices and these functions may have some properties (such as increasing differences) in particular components. Increasing differences for particular components is formalized as follows. Consider posets $X_1, \ldots, X_N$, and let $X = X_1 \times \cdots \times X_N$. For $x = (x_1, \ldots, x_N) \in X$, and for $i \neq j$, let

$$x_{-(i,j)} = (x_1, \ldots, x_{i-1}, x_{i+1}, \ldots, x_{j-1}, x_{j+1}, \ldots, x_N)$$

be the $(N-2)$-tuple formed by removing components x_i and x_j, and let

$$X_{-(i,j)} = X_1 \times \cdots \times X_{i-1} \times X_{i+1} \times \cdots \times X_{j-1} \times X_{j+1} \times \cdots \times X_N$$

be the $(N-2)$-fold Cartesian product formed by removing posets X_i and X_j, and denote the values of $f : X \to \mathbb{R}$ by $f(x_i, x_j, x_{-(i,j)})$. A function $f : X \to \mathbb{R}$ has ***increasing differences in*** (x_i, x_j), for $i \neq j$, if for every $x_{-(i,j)} \in X_{-(i,j)}$, $f(x_i, x_j, x_{-(i,j)})$ has increasing differences in (x_i, x_j). A function $f : X \to \mathbb{R}$ has ***increasing differences on*** X, if for every $i \neq j$, f has increasing differences in (x_i, x_j). This notation is helpful to prove relations between increasing differences and supermodular function below.

In a lattice game, $(X_i, u_i)_{i=1}^I$, the notion that player i tends to take higher actions when their opponents take higher actions is sometimes formalized by requiring that payoff of player i has increasing differences in (x_i, x_{-i}).

Examples 3.1 Consider the following examples.

1. In the coordination game (Example 1.1, page 2), the payoff of each player has increasing differences. If technology A is lower than technology B, denoted $A \prec B$, then $u_1(B, A) - u_1(A, A) = -2 \leq 1 = u_1(B, B) - u_1(A, B)$. If technology B is lower than technology A, denoted $B \prec A$, then $u_1(A, B) - u_1(B, B) = -1 \leq 2 = u_1(A, A) - u_1(B, A)$. Similarly, player 2 payoff has increasing differences as well.
2. In dove hawk game (Example 1.2, page 3), the payoff of each player does not have increasing differences. If $D \prec H$, then $u_1(H, D) - u_1(D, D) = 1 \not\leq -1 = u_1(H, H) - u_1(D, H)$. If $H \prec D$, then $u_1(D, H) - u_1(H, H) = 1 \not\leq -1 = u_1(D, D) - u_1(H, D)$. Similarly, player 2 payoff does not have increasing differences.
3. In matching pennies (Example 1.3, page 4), player 2 payoff has increasing differences, and player 1 payoff does not. If $H \prec T$, then $u_1(T, H) - u_1(H, H) = 2 \not\leq -2 = u_1(T, T) - u_1(H, T)$, and $u_2(T, H) - u_2(H, H) = -2 \leq 2 = u_2(T, T) - u_2(H, T)$. If $T \prec H$, then $u_1(H, T) - u_1(T, T) = 2 \not\leq -2 = u_1(H, H) - u_1(T, H)$, and $u_2(H, T) - u_2(T, T) = -2 \leq 2 = u_2(H, H) - u_2(T, H)$.
4. In the team trust game (Example 1.4, page 5), payoff differences are constant for both players, and therefore, payoff for each player satisfies definition of increasing differences.
5. In the multi-player coordination game (Example 1.5, page 7), each player i has increasing differences, because

$$u_i(1, x_{-i}) - u_i(0, x_{-i}) = \frac{2}{I - 1}\left(\sum_{j \neq i} x_j\right) - 1,$$

and this is (weakly) increasing in x_{-i}.
6. In the multi-player externality game (Example 1.6, page 7), no player i has increasing differences, because when x_{-i} increases from the zero vector 0 to a nonzero vector x_{-i}, $u_i(1, 0) - u_i(0, 0) = 1 - c \not\leq 1 - c - 1 = u_i(1, x_{-i}) - u_i(0, x_{-i})$. (The difference $u_i(1, x_{-i}) - u_i(0, x_{-i})$ is constant at $-c$ when evaluated at two different nonzero x_{-i}.)

3.1.2 Supermodular Function

Another way in which incentives to take higher actions can be formalized is by supermodular functions, which encapsulate increasing differences.

Let $(X, \preceq)$ be a lattice. A function $f : X \to \mathbb{R}$ is ***supermodular on*** X, if for every $x, y \in X$, $f(x) - f(x \wedge y) \leq f(x \vee y) - f(y)$. If x and y are comparable, the condition holds trivially, and therefore, a nontrivial application requires x and y to be noncomparable. Therefore, if X is a chain, then every function $f : X \to \mathbb{R}$ is supermodular. In particular, every function $f : \mathbb{R} \to \mathbb{R}$ is supermodular.

The supermodular property for particular components of a product lattice is formalized as follows. Consider lattices $X_1, \ldots, X_N$, and let $X = X_1 \times \cdots \times X_N$ be the product lattice. For $x = (x_1, \ldots, x_N) \in X$, and for a fixed i, let $x_{-i} = (x_1, \ldots, x_{i-1}, x_{i+1}, \ldots, x_N)$ be the $(N-1)$-tuple formed by removing x_i, and let $X_{-i} = X_1 \times \cdots \times X_{i-1} \times X_{i+1} \times \cdots \times X_N$ be the $(N-1)$-fold Cartesian product formed by removing X_i, and denote the values of $f : X \to \mathbb{R}$ by $f(x_i, x_{-i})$. For $i = 1, \ldots, N$, a function $f : X \to \mathbb{R}$ is ***supermodular on*** X_i, if for every $x_{-i} \in X_{-i}$, $f(\cdot, x_{-i})$ is supermodular on X_i. This notation is helpful to prove relations between increasing differences property and supermodular function given below.

The following relations between increasing differences and supermodular function are due to Topkis (1978). The proof is expanded to include additional detail. The product notation above and additional notation in the proof provide intermediate steps to see the logic more clearly.

Theorem 3.2 *Let $X_1, \ldots, X_N$ be lattices, $X = X_1 \times \cdots \times X_N$ be the product lattice, and $f : X \to \mathbb{R}$.*

1. *If f is supermodular on X, then f has increasing differences on X.*
2. *If f has increasing differences on X, and for every i, f is supermodular on X_i, then f is supermodular on X.*
3. *Suppose $X_1, \ldots, X_N$ are chains.*
   f is supermodular on X, if, and only if, f has increasing differences on X.

Proof For statement (1), suppose f is supermodular on X. Fix $i \neq j$, consider $\hat{x}_i \preceq \tilde{x}_i$ and $\hat{x}_j \preceq \tilde{x}_j$, and let $y = (\tilde{x}_i, \hat{x}_j, x_{-(i,j)})$ and $z = (\hat{x}_i, \tilde{x}_j, x_{-(i,j)})$. Then

$$\begin{aligned} f(\tilde{x}_i, \hat{x}_j, x_{-(i,j)}) - f(\hat{x}_i, \hat{x}_j, x_{-(i,j)}) &= f(y) - f(y \wedge z) \\ &\leq f(y \vee z) - f(z) \\ &= f(\tilde{x}_i, \tilde{x}_j, x_{-(i,j)}) - f(\hat{x}_i, \tilde{x}_j, x_{-(i,j)}), \end{aligned}$$

where the inequality holds because f is supermodular on X. As i, j are arbitrary, this shows that f satisfies increasing differences on X.

For statement (2), for $i = 1, \ldots, N+1$, let $z^i \in X$ be given by

$$z^i_j = \begin{cases} x_j & \text{if } j < i \\ x_j \wedge y_j & \text{if } j \geq i. \end{cases}$$

Notice that $z^1 = x \wedge y$ and $z^{N+1} = x$. Moreover, for $i = 1, \ldots, N$, let $\tilde{v}^i, \hat{v}^i \in X$ be given by

$$\tilde{v}^i_j = \begin{cases} x_j \vee y_j & \text{if } j < i \\ x_i & \text{if } j = i \\ y_j & \text{if } j > i, \end{cases} \qquad \hat{v}^i_j = \begin{cases} x_j \vee y_j & \text{if } j < i \\ x_i \wedge y_i & \text{if } j = i \\ y_j & \text{if } j > i, \end{cases}$$

and let $\tilde{w}^i, \hat{w}^i \in X$ be given by

$$\tilde{w}^i_j = \begin{cases} x_j \vee y_j & \text{if } j < i \\ x_i \vee y_i & \text{if } j = i \\ y_j & \text{if } j > i, \end{cases} \qquad \hat{w}^i_j = \begin{cases} x_j \vee y_j & \text{if } j < i \\ y_i & \text{if } j = i \\ y_j & \text{if } j > i. \end{cases}$$

Notice that $\hat{w}^1 = y$, $\tilde{w}^N = x \vee y$, and for $1 \leq i \leq N-1$, $\tilde{w}^i = \hat{w}^{i+1}$. This shows that

$$\begin{aligned} f(x) - f(x \wedge y) &= \textstyle\sum_{i=1}^N [f(z^{i+1}) - f(z^i)] \\ &\leq \textstyle\sum_{i=1}^N [f(\tilde{v}^i) - f(\hat{v}^i)] \\ &\leq \textstyle\sum_{i=1}^N [f(\tilde{w}^i) - f(\hat{w}^i)] \\ &= f(x \vee y) - f(y), \end{aligned}$$

where the first inequality follows from increasing differences on X and the second inequality follows from f is supermodular on X_i, for every i. Therefore, f is supermodular on X.

For statement (3), sufficiency follows from statement (1) and necessity follows from statement (2), because every function on a chain is supermodular, and therefore, for every i, f is supermodular on X_i. □

In statement (2), the condition that f is supermodular on X_i cannot be dropped in general, as shown by the following example.

Examples 3.3 Let $X_1 = \{a, b, c, d\}$ with $a \prec b \prec d, a \prec c \prec d$, and b and c are not comparable. Let $X_2 = \{0, 1\}$ with $0 \prec 1$ and consider $X = X_1 \times X_2$ with the product order. Let $f : X \to \mathbb{R}$ be given by

$$f(a, 0) = 1,\ f(b, 0) = 4,\ f(c, 0) = 2,\ f(d, 0) = 3,$$
$$f(a, 1) = 1,\ f(b, 1) = 5,\ f(c, 1) = 3,\ f(d, 1) = 6.$$

Then f has increasing differences in (x_1, x_2) and in (x_2, x_1), but f is not supermodular on X, because if we let $y = (b, 0)$ and $z = (c, 0)$, then $f(y) - f(y \wedge z) = 3 \not\leq 1 = f(y \vee z) - f(z)$. The conclusion does not hold, because $f(\cdot, 0)$ is not supermodular on X_1. Also, $f(\cdot, 1)$ is not supermodular on X_1.

When a function is twice continuously differentiable, a convenient tool to check for increasing differences and supermodularity is available. In the following, an open sublattice in $\mathbb{R}^n$ is a sublattice that is an open set as well.

Theorem 3.4 *Let X, Y be open sublattices of $\mathbb{R}$ and $f : X \times Y \to \mathbb{R}$ be twice continuously differentiable.*
f has increasing differences on $X \times Y$, if, and only if, $\frac{\partial^2 f}{\partial y \partial x} \geq 0$ on $X \times Y$.

Proof From the definition, it follows that f has increasing differences on $X \times Y$, if, and only, if for every $\tilde{x} > \hat{x}$, $f(\tilde{x}, y) - f(\hat{x}, y)$ is (weakly) increasing in y, which is true, if, and only if for every $\tilde{x} > \hat{x}$, $\frac{f(\tilde{x},y)-f(\hat{x},y)}{\tilde{x}-\hat{x}}$ is (weakly) increasing in y, and this is equivalent to for every $\tilde{x} > \hat{x}$, $\frac{\partial}{\partial y}\frac{f(\tilde{x},y)-f(\hat{x},y)}{\tilde{x}-\hat{x}} \geq 0$. For sufficiency, the conclusion follows from

$$\frac{\partial}{\partial y}\frac{\partial f}{\partial x} = \frac{\partial}{\partial y} \lim_{\tilde{x} \to \hat{x}} \frac{f(\tilde{x}, y) - f(\hat{x}, y)}{\tilde{x} - \hat{x}} = \lim_{\tilde{x} \to \hat{x}} \frac{\partial}{\partial y} \frac{f(\tilde{x}, y) - f(\hat{x}, y)}{\tilde{x} - \hat{x}} \geq 0,$$

where the last equality uses the assumption that f is twice continuously differentiable. Necessity follows from

$$\frac{\partial}{\partial y}(f(\tilde{x}, y) - f(\hat{x}, y)) = \int_{\hat{x}}^{\tilde{x}} \frac{\partial}{\partial x}\frac{\partial f}{\partial y}(x, y)\, dx \geq 0.$$

□

Combining this with the earlier result yields the following theorem.

Theorem 3.5 *Let* $X = X_1 \times \cdots \times X_N$ *be an open sublattice of* $\mathbb{R}^N$ *and* $f : X \to \mathbb{R}$ *is twice continuously differentiable.*
The following are equivalent.

1. f *is supermodular on* X.
2. f *has increasing differences on* X.
3. *For every* $i, j = 1, \ldots, N, i \neq j$, $\frac{\partial^2 f}{\partial x_i \partial x_j} \geq 0$ *on* $X_i \times X_j$.

Proof The equivalence of (1) and (2) is proved in statement (3) of Theorem 3.2 (page 50). The equivalence of (2) and (3) follows from Theorem 3.4 (page 52). □

This characterization provides a convenient tool to check when a function is supermodular or satisfies increasing differences.

Examples 3.6 Consider the following examples.

1. The Cobb–Douglas function $f : \mathbb{R}^N_+ \to \mathbb{R}$ given by $f(x) = x_1^{\alpha_1} x_2^{\alpha_2} \cdots x_N^{\alpha_N}$, where $\alpha_n > 0$ for every n, is supermodular. This follows because each $x_n \geq 0$ and for every $i, j = 1 \ldots, N$ with $i \neq j$,

$$\frac{\partial^2 f}{\partial x_i \partial x_j} = \alpha_i \alpha_j x_i^{\alpha_i - 1} x_j^{\alpha_j - 1} \prod_{k \neq i,j} x_k^{\alpha_k} \geq 0.$$

2. The Leontief perfect complements function given by $f(x_1, \ldots, x_N) = \min\{x_1, \ldots, x_N\}$ is supermodular. This can be seen as follows. For notational convenience, let $\min_n\{x_n\}$ denote $\min\{x_1, \ldots, x_N\}$ and notice that for every $x, y \in \mathbb{R}^N$, $\min_n\{x_n\} \wedge \min_n\{y_n\} = \min_n\{x_n \wedge y_n\}$ and $\min_n\{x_n\} \vee \min_n\{y_n\} \leq \min_n\{x_n \vee y_n\}$. Therefore, for every $x, y \in \mathbb{R}^N$,

$$\begin{aligned} f(x) + f(y) &= \min_n\{x_n\} + \min_n\{y_n\} \\ &= \min_n\{x_n\} \wedge \min_n\{y_n\} + \min_n\{x_n\} \vee \min_n\{y_n\} \\ &\leq \min_n\{x_n \wedge y_n\} + \min_n\{x_n \vee y_n\} \\ &= f(x \wedge y) + f(x \vee y). \end{aligned}$$

3. In a two player lattice game (X_1, X_2, u_1, u_2), a useful construction to include codirectional incentives is to think of payoff as a

sum of personal effect and interactive effect. For example, suppose $u_1(x_1, x_2) = f(x_1) + \phi(x_1, x_2)$ where f captures personal effect of player 1 action and ϕ captures interactive effect of actions of both players on player 1 payoff. It is immediate that u_1 has increasing differences in (x_1, x_2), if, and only if, $\phi(x_1, x_2)$ has increasing differences in (x_1, x_2). One way in which this is achieved is when ϕ has a multiplicatively separable form, that is, $\phi(x_1, x_2) = g(x_1)h(x_2)$, where both g and h are (weakly) increasing functions. In particular, if X_1 and X_2 are open sublattices of $\mathbb{R}$ and u_1 is twice continuously differentiable, then

$$\frac{\partial^2 u_1}{\partial x_2 \partial x_1} = g'(x_1)h'(x_2) \geq 0.$$

Examples of this formulation include the following forms useful in applications: $\phi(x_1, x_2) = x_1 x_2$, $\phi(x_1, x_2) = \sqrt{x_1 x_2}$, $\phi(x_1, x_2) = x_1\sqrt{x_2}$, $\phi(x_1, x_2) = x_1 \ln x_2$, $\phi(x_1, x_2) = -x_1 e^{-x_2}$, and $\phi(x_1, x_2) = A x_1^\alpha x_2^\beta$ for $A, \alpha, \beta > 0$, and where $x_1, x_2 \geq 0$, as needed. These generalize to additional cases and to more than two players in a natural manner.

4. The function f on pairs of strictly positive real numbers $x > 0$ and $y > 0$ given by $f(x_1, x_2) = (x_1 + x_2)^2$ is supermodular, because $\frac{\partial^2 f}{\partial x_2 \partial x_1} = 2$. Its monotonic transformation given by $h(x_1, x_2) = 2\ln(x_1 + x_2)$ is not supermodular, because $\frac{\partial^2 h}{\partial x_2 \partial x_1} = -\frac{2}{(x_1+x_2)^2} < 0$.
5. In Bertrand oligopoly (Example 1.7, page 8), profit of firm i from setting price x_i when opponents set prices x_{-i} is given by

$$u_i(x_i, x_{-i}) = (x_i - c)\left(a - bx_i + \beta\left(\sum_{j \neq i} x_j\right)\right).$$

For every firm i and for every competitor $j \neq i$, $\frac{\partial^2 u_i}{\partial x_j \partial x_i} = \beta > 0$. Therefore, this payoff function is supermodular (and satisfies increasing differences).
6. In Cournot oligopoly (Example 1.8, page 9), profit of firm i from producing quantity x_i when opponents produce quantities x_{-i} is given by

$$u_i(x_i, x_{-i}) = \left(a - b\left(\sum_{i=1}^{I} x_i\right)\right) x_i - cx_i.$$

For every firm i and for every competitor $j \neq i$, $\frac{\partial^2 u_i}{\partial x_j \partial x_i} = -b < 0$. Therefore, this payoff function is not supermodular (and does not satisfy increasing differences).

3.1.3 Single Crossing Property

For posets $(X, \preceq_X)$ and $(Y, \preceq_Y)$, a function $f : X \times Y \to \mathbb{R}$ has ***single crossing property in*** (x, y), if for every $\hat{x} \preceq \tilde{x}$ and for every $\hat{y} \preceq \tilde{y}$, (1) $f(\hat{x}, \hat{y}) \leq f(\tilde{x}, \hat{y}) \Rightarrow f(\hat{x}, \tilde{y}) \leq f(\tilde{x}, \tilde{y})$ and (2) $f(\hat{x}, \hat{y}) < f(\tilde{x}, \hat{y}) \Rightarrow f(\hat{x}, \tilde{y}) < f(\tilde{x}, \tilde{y})$.

Single crossing property requires that for each $\hat{x} \preceq \tilde{x}$, if the function $f(\tilde{x}, \cdot) - f(\hat{x}, \cdot)$ crosses zero (from below) at $\hat{y}$, that is, $f(\tilde{x}, \hat{y}) - f(\hat{x}, \hat{y}) \geq 0$, then it can never go below zero again, that is, for every $\tilde{y} \succeq \hat{y}$, $f(\tilde{x}, \tilde{y}) - f(\hat{x}, \tilde{y}) \geq 0$. Moreover, if the function $f(\tilde{x}, \cdot) - f(\hat{x}, \cdot)$ strictly crosses zero once, it can never go back to zero (or below zero) again.

Single crossing property in not commutative in its components, that is, f has single crossing in (x, y) does not necessarily imply that f has single crossing property in (y, x).

Example 3.7 Consider posets $X = Y = \{0, 1\}$ with $0 \prec 1$, and $f : X \times Y \to \mathbb{R}$ is given by $f(0, 0) = 1$, $f(1, 0) = 4$, $f(0, 1) = 2$, $f(1, 1) = 3$. Then f has single crossing property in (x, y) but not in (y, x).

It is easy to check that if f has increasing differences in (x, y), then f has single crossing property in (x, y). The converse is not necessarily true. The previous example shows this as well, because in the example, f has single crossing property in (x, y), but f does not have increasing differences in (x, y).

It follows immediately from the definition that single crossing property is invariant to strictly increasing transformations, that is, if f has single crossing property in (x, y) and $g : \mathbb{R} \to \mathbb{R}$ is a strictly increasing function, then $g \circ f$ has single crossing property in (x, y). In this sense, single crossing property is an ordinal generalization of increasing differences.

Single crossing property for particular components of a product space is formalized as follows. Consider posets $X_1, \ldots, X_N$, and let $X = X_1 \times \cdots \times X_N$. As earlier, for $x = (x_1, \ldots, x_N) \in X$, and for $i \neq j$, let $x_{-(i,j)}$ be the $(N-2)$-tuple formed by removing components x_i and x_j, and let $X_{-(i,j)}$ be the $(N-2)$-fold Cartesian product formed by removing posets X_i and X_j, and denote the values of $f : X \to \mathbb{R}$ by $f(x_i, x_j, x_{-(i,j)})$. A function

$f : X \to \mathbb{R}$ has ***single crossing property in*** (x_i, x_j), for $i \neq j$, if for every $x_{-(i,j)} \in X_{-(i,j)}$, $f(x_i, x_j, x_{-(i,j)})$ has single crossing property in (x_i, x_j). A function $f : X \to \mathbb{R}$ has ***single crossing property on*** X, if for every i, j with $i \neq j$, f has single crossing property in (x_i, x_j).

3.1.4 *Quasisupermodular Function*

Let $(X, \preceq_X)$ be a lattice. A function $f : X \to \mathbb{R}$ is ***quasisupermodular on*** X, if for every $x, y \in X$, (1) $f(x \wedge y) \leq f(x) \Rightarrow f(y) \leq f(x \vee y)$, and (2) $f(x \wedge y) < f(x) \Rightarrow f(y) < f(x \vee y)$. If x and y are comparable, the condition holds trivially, and therefore, a nontrivial application requires x and y to be noncomparable. Therefore, if X is a chain, then every function $f : X \to \mathbb{R}$ is quasisupermodular. In particular, every function $f : \mathbb{R} \to \mathbb{R}$ is quasisupermodular.

It follows immediately from the definition that quasisupermodular property is invariant to strictly increasing transformations, that is, if f is quasisupermodular on X and $g : \mathbb{R} \to \mathbb{R}$ is a strictly increasing function, then $g \circ f$ is quasisupermodular on X. In this sense, quasisupermodular property is an ordinal generalization of supermodular property.

It is easy to check that if f is supermodular on X, then f is quasisupermodular on X. The converse is not necessarily true.

Example 3.8 The function h on pairs of strictly positive real numbers $x > 0$ and $y > 0$ given by $h(x, y) = 2\ln(x + y)$ is a quasisupermodular function, because it is a strictly increasing transformation of the supermodular function $f(x, y) = (x + y)^2$, but h is not supermodular, because $\frac{\partial^2 h}{\partial x_2 \partial x_1} = -\frac{2}{(x_1+x_2)^2} < 0$.

Theorem 3.9 *Let* $X_1, \ldots, X_N$ *be lattices,* $X = X_1 \times \cdots \times X_N$, *and* $f : X \to \mathbb{R}$.
If f *is quasisupermodular on* X, *then* f *has single crossing property on* X.

Proof Suppose f is quasisupermodular on X. Fix $i, j \in \{1, \ldots, N\}$ with $i \neq j$, and $x_{-(i,j)} \in X_{-(i,j)}$. Consider $\hat{x}_i \preceq \tilde{x}_i$ and $\hat{x}_j \preceq \tilde{x}_j$. Let $y = (\tilde{x}_i, \hat{x}_j, x_{-(i,j)})$ and $z = (\hat{x}_i, \tilde{x}_j, x_{-(i,j)})$. Then

$$\begin{aligned} & f(\tilde{x}_i, \hat{x}_j, x_{-(i,j)}) \geq f(\hat{x}_i, \hat{x}_j, x_{-(i,j)}) \\ \Leftrightarrow \quad & f(y) \geq f(y \wedge z) \\ \Rightarrow \quad & f(y \vee z) \geq f(z) \\ \Leftrightarrow \quad & f(\tilde{x}_i, \tilde{x}_j, x_{-(i,j)}) \geq f(\hat{x}_i, \tilde{x}_j, x_{-(i,j)}), \end{aligned}$$

where the implication follows from f is quasisupermodular on X. This shows that for every i, j with $i \neq j$, f has single crossing property in (x_i, x_j). □

In general, single crossing property (even on all pairs of components) does not imply quasisupermodular property, as shown by the following example.

Example 3.10 Let $X_1 = \{a, b, c, d\}$ with $a \prec b \prec d, a \prec c \prec d$, and b and c are not comparable. Let $X_2 = \{0, 1\}$ with $0 \prec 1$ and consider $X = X_1 \times X_2$ with the product partial order. Let $f : X \to \mathbb{R}$ be given by

$$f(a, 0) = 1, f(b, 0) = 2, f(c, 0) = 4, f(d, 0) = 3,$$
$$f(a, 1) = 1, f(b, 1) = 2, f(c, 1) = 4, f(d, 1) = 3.$$

It is easy to check that f has single crossing property in (x_1, x_2) and in (x_2, x_1), but f is not quasisupermodular on X, because if we let $y = (b, 0)$ and $z = (c, 0)$, then $f(y) - f(y \wedge z) = 1 \geq 0$, but $f(y \vee z) - f(z) = -1 \ngeq 0$.

3.1.5 *Increasing Maximizers*

In order to understand the behavior of best responses with codirectional incentives, it is useful to consider the more general problem of constrained optimization with parameters. Let X, T be sets, S a subset of X, $f : X \times T \to \mathbb{R}$, and consider the maximization problem

$$\max_{x \in S} f(x, t).$$

Here, x is the decision variable in the decision space X, t is a parameter in the parameter space T, S is a constraint on the available decision choices, and f is the payoff. The object of interest is the solution to this problem, denoted

$$B(S, t) = \arg\max_{x \in S} f(x, t).$$

Elements of $B(S, t)$ are the payoff maximizing choices when the parameter is t and the decision-making constraint is S. In other words, if the goal is to maximize payoff f when the decision-making environment is constrained by (S, t), then $B(S, t)$ is the set of best choices.

Understanding properties of $B(S, t)$ and how $B(S, t)$ changes with (S, t) is central to understanding the structure of best choices and how best choices change with the decision-making environment. This is important in game theory, because the best response set of each player i is a special case of this general problem.

Applying this terminology in a lattice game $\Gamma = (X_i, u_i)_{i=1}^I$, for each player i, let X_i be the decision space with decision variable denoted x_i, let X_{-i} be the parameter space with parameter denoted x_{-i}, let the constraint set be the whole decision space X_i, and let payoff be u_i. The maximization problem above specializes to the case of best response of player i to x_{-i},

$$B_i(x_{-i}) = \arg\max_{x_i \in X_i} u_i(x_i, x_{-i}),$$

where we suppress X_i in the more general notation $B_i(X_i, x_{-i})$.

Similarly, applying this terminology in a parameterized lattice game given by $\Gamma = ((X_i, u_i)_{i=1}^I, T)$, for each player i, let X_i be the decision space with decision variable denoted x_i, let $X_{-i} \times T$ be the parameter space with parameter denoted (x_{-i}, t), let the constraint set be the whole decision space X_i, and let payoff be u_i. The maximization problem above specializes to the case of best response of player i to x_{-i} at t,

$$B_i(x_{-i}, t) = \arg\max_{x_i \in X_i} u_i(x_i, x_{-i}, t),$$

where we suppress X_i in the more general notation $B_i(X_i, x_{-i}, t)$.

The motivation for codirectional incentives is to identify conditions on payoffs under which if opponents of player i take a higher action, it is in player i's best interest to take a higher action as well. That is, under which conditions is $B_i(x_{-i})$ (weakly) increasing in x_{-i}, or equivalently, when is it the case that $\hat{x}_{-i} \preceq \tilde{x}_{-i}$ implies $B_i(\hat{x}_{-i}) \sqsubseteq B_i(\tilde{x}_{-i})$? The increasing maximizers theorem below provides a general answer to this question.

The following characterization of increasing maximizers is due to Milgrom and Shannon (1994). Their proof is expanded to include more detail to make it more accessible. Let X be a lattice, T a poset, S a subset of X, and $f : X \times T \to \mathbb{R}$. Consider the maximization problem $\max_{x \in S} f(x, t)$ and let

$$B(S, t) = \arg\max_{x \in S} f(x, t)$$

be the solution to the maximization problem. The solution set $B(S, t)$ is ***(weakly) increasing in*** (S, t), if for every $\hat{S} \sqsubseteq \tilde{S}$ and for every $\hat{t} \preceq \tilde{t}$, $B(\hat{S}, \hat{t}) \sqsubseteq B(\tilde{S}, \tilde{t})$. Strongly increasing maximizers are discussed in Shannon (1995).

Theorem 3.11 (Increasing Maximizers Theorem) *Let X be a lattice, T a poset, S a subset of X, $f : X \times T \to \mathbb{R}$, and $B(S, t) = \arg\max_{x \in S} f(x, t)$. $B(S, t)$ is (weakly) increasing in (S, t), if, and only if, f has single crossing property in (x, t) and for every $t \in T$, $f(\cdot, t)$ is quasisupermodular on X.*

Proof For sufficiency, suppose $B(S, t)$ is (weakly) increasing in (S, t). To show that f has single crossing property in (x, t), consider $\hat{x} \preceq \tilde{x}$ and $\hat{t} \preceq \tilde{t}$. Suppose $f(\hat{x}, \hat{t}) \le f(\tilde{x}, \hat{t})$. If $\hat{x} = \tilde{x}$ then $f(\hat{x}, \tilde{t}) \le f(\tilde{x}, \tilde{t})$ is trivially true, so suppose $\hat{x} \prec \tilde{x}$, and let $S = \{\hat{x}, \tilde{x}\}$. In this case, $\tilde{x} \in B(S, \hat{t})$, and combined with $B(S, \hat{t}) \sqsubseteq B(S, \tilde{t})$, it follows that $\tilde{x} \in B(S, \tilde{t})$, and therefore, $f(\hat{x}, \tilde{t}) \le f(\tilde{x}, \tilde{t})$, as desired. Now suppose $f(\hat{x}, \hat{t}) < f(\tilde{x}, \hat{t})$, which implies that $\hat{x} \prec \tilde{x}$, and let $S = \{\hat{x}, \tilde{x}\}$. In this case, $\{\tilde{x}\} = B(S, \hat{t})$, and combined with $B(S, \hat{t}) \sqsubseteq B(S, \tilde{t})$, it follows that $B(S, \tilde{t}) = \{\tilde{x}\}$, and therefore, $f(\hat{x}, \tilde{t}) < f(\tilde{x}, \tilde{t})$.

To show that for every $t \in T$, $f(\cdot, t)$ is quasisupermodular on X, fix $t \in T$ and consider $x, y \in X$. Suppose $f(x, t) \ge f(x \wedge y, t)$. Let $\hat{S} = \{x, x \wedge y\}$ and $\tilde{S} = \{y, x \vee y\}$, and notice that $\hat{S} \sqsubseteq \tilde{S}$. In this case, $x \in B(\hat{S}, t)$, and combined with $B(\hat{S}, t) \sqsubseteq B(\tilde{S}, t)$, it follows that $x \vee y \in B(\tilde{S}, t)$, and therefore, $f(x \vee y, t) \ge f(y, t)$, as desired. Now suppose $f(x, t) > f(x \wedge y, t)$. In this case, $\{x\} = B(\hat{S}, t)$, and combined with $B(\hat{S}, t) \sqsubseteq B(\tilde{S}, t)$, it follows that $B(\tilde{S}, t) = \{x \vee y\}$, and therefore, $f(x \vee y, t) > f(y, t)$.

For necessity, suppose f has single crossing property in (x, t) and for every $t \in T$, $f(\cdot, t)$ is quasisupermodular on X, and consider $\hat{S} \sqsubseteq \tilde{S}$ and $\hat{t} \preceq \tilde{t}$. If either $B(\hat{S}, \hat{t})$ is empty or $B(\tilde{S}, \tilde{t})$ is empty, then the conclusion follows trivially. Otherwise, let $x \in B(\hat{S}, \hat{t})$ and $y \in B(\tilde{S}, \tilde{t})$. As $\hat{S} \sqsubseteq \tilde{S}$, it follows that $x \wedge y \in \hat{S}$ and $x \vee y \in \tilde{S}$, and therefore,

$$
\begin{aligned}
x \in B(\hat{S}, \hat{t}) \Rightarrow \quad & f(x, \hat{t}) \ge f(x \wedge y, \hat{t}) \\
\Rightarrow\; & f(x \vee y, \hat{t}) \ge f(y, \hat{t}) \\
\Rightarrow\; & f(x \vee y, \tilde{t}) \ge f(y, \tilde{t}),
\end{aligned}
$$

where the second implication follows from $f(\cdot, \hat{t})$ being quasisupermodular and the third implication follows from f having single crossing property in (x, t). Combined with $y \in B(\tilde{S}, \tilde{t})$ and $x \vee y \in \tilde{S}$, it follows that $x \vee y \in$

$B(\tilde{S}, \tilde{t})$. Similarly,

$$\begin{aligned} y \in B(\tilde{S}, \tilde{t}) \Rightarrow \quad & f(y, \tilde{t}) \geq f(x \vee y, \tilde{t}) \\ \Rightarrow \quad & f(y, \hat{t}) \geq f(x \vee y, \hat{t}) \\ \Rightarrow & f(x \wedge y, \hat{t}) \geq f(x, \hat{t}), \end{aligned}$$

where the second implication follows from f having single crossing property in (x, t) and the third implication follows from $f(\cdot, \hat{t})$ being quasisupermodular. Combined with $x \in B(\hat{S}, \hat{t})$ and $x \wedge y \in \hat{S}$, it follows that $x \wedge y \in B(\hat{S}, \hat{t})$. We conclude that $B(\hat{S}, \hat{t}) \sqsubseteq B(\tilde{S}, \tilde{t})$. □

The increasing maximizers theorem gives sufficient conditions for best response of player i to be increasing in actions of opponents. In particular, if u_i has single crossing property in (x_i, x_{-i}) and for each x_{-i}, $u_i(\cdot, x_{-i})$ is quasisupermodular on X_i, then $B_i(x_{-i})$ is (weakly) increasing in x_{-i}. In other words, for every $\hat{x}_{-i}, \tilde{x}_{-i} \in X_{-i}$, $\hat{x}_{-i} \preceq \tilde{x}_{-i}$ implies $B_i(\hat{x}_{-i}) \sqsubseteq B_i(\tilde{x}_{-i})$.

The increasing maximizers theorem has the following useful corollaries.

Corollary 3.12 *If f has increasing differences in (x, t) and for every $t \in T$, $f(\cdot, t)$ is supermodular on X, then $B(S, t)$ is (weakly) increasing in (S, t).*

Proof As increasing differences in (x, t) imply single crossing property in (x, t) and a supermodular function is quasisupermodular, the result follows from the previous theorem. □

Corollary 3.13 *Let X be a chain, T a poset, S a subset of X, $f : X \times T \to \mathbb{R}$, and $B(S, t) = \arg\max_{x \in S} f(x, t)$.*
$B(S, t)$ is (weakly) increasing in (S, t), if, and only if, f has single crossing property in (x, t).

Proof When X is a chain, every function on X is quasisupermodular, and therefore, the condition *for every $t \in T$, $f(\cdot, t)$ is quasisupermodular on X* in the previous theorem is satisfied automatically. □

Corollary 3.14 *Let X be a lattice, S a subset of X, $f : X \to \mathbb{R}$, and $B(S) = \arg\max_{x \in S} f(x)$.*
$B(S)$ is (weakly) increasing in S, if, and only if, f is quasisupermodular on X.

Proof Apply Theorem 3.11 (page 59) with poset $T = \{0\}$. □

Corollary 3.15 *Let X be a chain and S a subset of X.*
For every $f : X \to \mathbb{R}$, $B(S) = \arg\max_{x \in S} f(x)$ is (weakly) increasing in S.

Proof Apply the previous corollary and note that every function $f : X \to \mathbb{R}$ on a chain X is quasisupermodular on X. □

Corollary 3.16 *Let X be a lattice, $f : X \to \mathbb{R}$, and $B(S) = \arg\max_{x \in S} f(x)$. If f is quasisupermodular on X and S is a sublattice of X, then $B(S)$ is a sublattice of X.*

Proof As S is a sublattice of X, $S \sqsubseteq S$, and therefore, the previous corollary yields $B(S) \sqsubseteq B(S)$, which is true, if, and only if, $B(S)$ is a sublattice of X.□

3.2 Game with Strategic Complements

3.2.1 *Definition*

A game with strategic complements models a decentralized and interdependent decision-making environment in which each player i has codirectional incentives, that is, each player i has an incentive to take a higher action when opponents of i take a higher action.

One way to incorporate codirectional incentives is to posit that marginal benefit to player i from a higher action is greater when opponents take higher actions. This is formalized by u_i has increasing differences in (x_i, x_{-i}).

Single crossing property is weaker than increasing differences, and increasing maximizers theorem shows that single crossing property (combined with quasisupermodular property) is sufficient for increasing best responses. In other words, a sufficient condition for player i to have codirectional incentives is for player i payoff to have single crossing property in (x_i, x_{-i}) and quasisupermodular property in own action.

Strategic complements are formalized as follows. In a lattice game $\Gamma = (X_i, u_i)_{i=1}^I$, ***player i has strategic complements***, if u_i has single crossing property in (x_i, x_{-i}) and for each x_{-i}, $u_i(\cdot, x_{-i})$ is quasisupermodular on X_i.

A ***game with strategic complements***, or ***GSC***, is a lattice game in which every player i has strategic complements. In other words, a GSC is a lattice game $\Gamma = (X_i, u_i)_{i=1}^I$, where

a. There are finitely many players, indexed by $i \in \{1, \ldots, I\}$.
b. For each i, the action space of player i is X_i, where X_i is a nonempty, subcomplete lattice in $\mathbb{R}^{n_i}$ with the Euclidean order.
c. For each i, payoff for player i is $u_i : X_i \times X_{-i} \to \mathbb{R}$, where

   (a) u_i is upper semicontinuous on X and continuous on X_{-i}, and
   (b) u_i has single crossing property in (x_i, x_{-i}) and for each x_{-i}, $u_i(\cdot, x_{-i})$ is quasisupermodular on X_i.

Games with strategic complements have been studied by many authors. Early work includes that of Topkis (1978) and Topkis (1979). Economic applications using these ideas in two player games are available in Fudenberg and Tirole (1984) and Bulow et al. (1985). More general results are available in Lippman et al. (1987), Sobel (1988), Milgrom and Roberts (1990), Vives (1990), Zhou (1994), Echenique (2002), Echenique and Sabarwal (2003), and Amir et al. (2008), among others. Book length treatments are Topkis (1998) and Vives (1999). Models with aggregative games are considered in Jensen (2010) and Acemoglu and Jensen (2013).

Although a player's action space in a GSC is assumed to be finite dimensional, the material on codirectional incentives in the previous section is developed in its natural and more general setting. Moreover, proofs of results in this section are developed in a manner that shows their natural extension to more general settings with additional assumptions.

The coordination game, team trust game, multi-player coordination game, and Bertrand oligopoly are all games with strategic complements. As shown in Examples 3.1 (page 49), in each of these cases, the payoff of each player has increasing differences, which implies that payoff of each player satisfies single crossing property. Moreover, in each of these cases, the action space of each player is a chain and therefore, each player's payoff is quasisupermodular on their own action space.

3.2.2 *Increasing Best Response*

Recall that in a lattice game best response set of player i to opponent actions x_{-i} is $B_i(x_{-i}) = \arg\max_{x_i \in X_i} u_i(x_i, x_{-i})$, the best response correspondence for player i is $B_i : X_{-i} \rightrightarrows X_i$, mapping x_{-i} to $B_i(x_{-i})$, and the joint best response correspondence is $B : X \rightrightarrows X$, given by $B(x) = B_1(x_{-1}) \times B_2(x_{-2}) \times \cdots \times B_I(x_{-I})$.

At the more general level of a lattice game, the individual and joint best response correspondence are nonempty valued, compact valued, and upper hemicontinuous (Theorem 2.6, page 32). The increasing maximizers theorem and characterization of subcomplete lattices in Theorem 2.14 (page 22) show that they are subcomplete valued and increasing as well.

Theorem 3.17 *In every lattice game* $\Gamma = (X_i, u_i)_{i=1}^{I}$,

1. *If player i has strategic complements, then best response correspondence for player i is nonempty valued, compact valued, subcomplete valued, upper hemicontinuous, and (weakly) increasing.*
2. *If the game is a GSC, then joint best response correspondence is nonempty valued, compact valued, subcomplete valued, upper hemicontinuous, and (weakly) increasing.*

Proof For statement (1), let B_i denote the best response correspondence for player i. Theorem 2.26 (page 32) shows that B_i is nonempty valued, compact valued, and upper hemicontinuous. When player i has strategic complements, Corollary 3.16 (page 61) shows that for every x_{-i}, $B_i(x_{-i})$ is a sublattice, and therefore, by Theorem 2.14 (page 22), $B_i(x_{-i})$ is subcomplete in X_i. Moreover, Theorem 3.11 (page 59) implies that B_i is (weakly) increasing. This proves statement (1).

For statement (2), let B denote the joint best response correspondence. Theorem 2.26 (page 32) shows that B is nonempty valued, compact valued, and upper hemicontinuous. Statement (1) implies that for every i, $B_i(x_{-i})$ is subcomplete in X_i, and therefore, for every $x \in X$, $B(x)$ is subcomplete. Moreover, for every i, B_i is (weakly) increasing in x_{-i}, and therefore, B is (weakly) increasing in x. □

As shown in Examples 2.27 (page 33), in the coordination game, team trust game, multi-player coordination game, and Bertrand oligopoly, the best response correspondence for each player is (weakly) increasing, and therefore, the joint best response correspondence in the game is (weakly) increasing.

An important application of increasing maximizers theorem is to provide sufficient conditions on payoffs for best response of each player to be (weakly) increasing in actions of opponents. These conditions are not necessary, as the following example shows.

Table 3.1 Non-necessity of single crossing property

		P2 0	P2 1	P2 2
P1	0	2, 2	0, 0	1, 1
P1	1	0, 0	3, 3	2, 2
P1	2	1, 1	2, 2	4, 4

Example 3.18 Consider a two player, three action game, with actions ordered $0 \prec 1 \prec 2$. Payoffs are given in the bimatrix in Table 3.1 (page 64). Best response for each player $i = 1, 2$ is given by $B_i(0) = \{0\}$, $B_i(1) = \{1\}$, $B_i(2) = \{2\}$. It is a (strictly) increasing function in opponent action. Nevertheless, for each player $i = 1, 2$, payoff for player i does not have single crossing property, either in (x_1, x_2) or in (x_2, x_1), because $u_i(2, 0) > u_i(1, 0) \not\Rightarrow u_i(2, 1) > u_i(1, 1)$.

Example 3.18 (page 63) is new and appears to be the first example in the literature of a game that shows both that single crossing property is not necessary for increasing best responses and also that the weaker condition of interval dominance order due to Quah and Strulovici (2009) is not necessary for increasing best responses. It shows both in the simple setting of a two player, three action game in which action spaces are chains.

A recurring theme in the theory of monotone games is the role played by some combinations of iterated joint best response correspondence starting from $\inf X$ and from $\sup X$. These turn out to have information useful to bound serially undominated strategies and eventual behavior of adaptive dynamics. Consequently, they help to characterize dominance solvability and global stability. The situation is simplest for GSC.

In a GSC, ***best response dynamic from*** $\inf X$ is the sequence $(\underline{x}^n)_{n=0}^{\infty}$ in X, where $\underline{x}^0 = \inf X$, and for $n \geq 0$, $\underline{x}^{n+1} = \underline{B}(\underline{x}^n)$, where $\underline{B}(\underline{x}^n) = \inf B(\underline{x}^n)$, and ***best response dynamic from*** $\sup X$ is the sequence $(\overline{x}^n)_{n=0}^{\infty}$ in X, where $\overline{x}^0 = \sup X$, and for $n \geq 0$, $\overline{x}^{n+1} = \overline{B}(\overline{x}^n)$, where $\overline{B}(\overline{x}^n) = \sup B(\overline{x}^n)$. Joint best response correspondence is subcomplete valued implies that for every $n \geq 0$, $\underline{B}(\underline{x}^n) \in B(\underline{x}^n)$ and $\overline{B}(\overline{x}^n) \in B(\overline{x}^n)$.

Monotone increasing best response for strategic complements implies that the best response dynamic from $\inf X$ is (weakly) increasing, the best response dynamic from $\sup X$ is (weakly) decreasing, and their corresponding terms and respective limits are ordered. Upper hemicontinuity shows that their limits are Nash equilibria.

Theorem 3.19 *In a GSC, let* $(\underline{x}^n)_{n=1}^{\infty}$ *and* $(\overline{x}^n)_{n=1}^{\infty}$ *be best response dynamics from* $\inf X$ *and* $\sup X$*, respectively.*

1. *For every* n, $\underline{x}^n \preceq \underline{x}^{n+1} \preceq \overline{x}^{n+1} \preceq \overline{x}^n$.
2. *Let* $\overline{x} = \bigvee_{n=0}^{\infty} \overline{x}^n$ *and* $\underline{x} = \bigwedge_{n=0}^{\infty} \underline{x}^n$. *Then* $\underline{x} \preceq \overline{x}$.
3. $\underline{x} \in B(\underline{x})$ *and* $\overline{x} \in B(\overline{x})$.

Proof For statement (1), use induction on n. For $n = 0$, $\underline{x}^0 \preceq \overline{x}^0$ and B is (weakly) increasing imply

$$\underline{x}^0 = \inf X \preceq \underline{x}^1 = \underline{B}(\underline{x}^0) \preceq \overline{B}(\overline{x}^0) = \overline{x}^1 \preceq \sup X = \overline{x}^0.$$

Suppose $\underline{x}^n \preceq \underline{x}^{n+1} \preceq \overline{x}^{n+1} \preceq \overline{x}^n$. Then $\underline{x}^n \preceq \underline{x}^{n+1}$ implies $B(\underline{x}^n) \sqsubseteq B(\underline{x}^{n+1})$, whence $\underline{x}^{n+1} \preceq \underline{x}^{n+2}$, and $\overline{x}^{n+1} \preceq \overline{x}^n$ implies $B(\overline{x}^{n+1}) \sqsubseteq B(\overline{x}^n)$, whence $\overline{x}^{n+2} \preceq \overline{x}^{n+1}$. Moreover, $\underline{x}^{n+1} \preceq \overline{x}^{n+1}$ implies $B(\underline{x}^{n+1}) \sqsubseteq B(\overline{x}^{n+1})$, whence $\underline{x}^{n+2} \preceq \overline{x}^{n+2}$. Consequently, $\underline{x}^{n+1} \preceq \underline{x}^{n+2} \preceq \overline{x}^{n+2} \preceq \overline{x}^{n+1}$.

For statement (2), statement (1) shows that $(\underline{x}^n)_{n=1}^{\infty}$ is a (weakly) increasing sequence and $(\overline{x}^n)_{n=1}^{\infty}$is a (weakly) decreasing sequence and both sequences are in a subcomplete lattice. Therefore, $\underline{x}$ and $\overline{x}$ exist and are well defined. As $\underline{x}^n \preceq \overline{x}^n$ for every n, it follows that $\underline{x} \preceq \overline{x}$.

For statement (3), let $y^n = \underline{x}^{n+1}$. Then $(\underline{x}^n)_{n=0}^{\infty}$ and $(y^n)_{n=0}^{\infty}$ are sequences in X, $\underline{x}^n \to \underline{x}$, $y^n \to \underline{x}$, and for every n, $y^n \in B(\underline{x}^n)$. As B is upper hemicontinuous, it follows that $\underline{x} \in B(\underline{x})$. Similarly, it can be shown that $\overline{x} \in B(\overline{x})$. □

This theorem shows that best response dynamic from $\inf X$ is a (weakly) increasing sequence in X, best response dynamic from $\sup X$ is a (weakly) decreasing sequence in X, and at every index along these two dynamics, the profile of play in the one from $\inf X$ is lower than its corresponding counterpart from $\sup X$.

Moreover, the limit of each of these best response dynamics is a fixed point of the joint best response correspondence, and therefore, a Nash equilibrium in the game. Consequently, both best response dynamics provide a constructive method to compute Nash equilibrium in the game. In particular, with finitely many actions, both dynamics necessarily converge to a Nash equilibrium in finitely many iterations.

3.2.3 *Nash Equilibrium Set*

In GSC, the Nash equilibrium set is a nonempty, complete lattice, as shown in Zhou (1994).

Theorem 3.20 *In every GSC, the Nash equilibrium set is a nonempty, complete lattice (in the relative partial order from X).*

Proof As the joint best response correspondence $B : X \rightrightarrows X$ is nonempty valued, subcomplete valued, and (weakly) increasing, Theorem 2.18 (page 25) shows that the set of fixed points of B is a nonempty complete lattice (in the relative partial order from X). □

This shows that in every GSC, there is at least one Nash equilibrium, and for every pair of Nash equilibria there is a Nash equilibrium that is the least upper bound of this pair and another that is the greatest lower bound of this pair. Moreover, with multiple equilibria, there is always a smallest Nash equilibrium in which everyone is playing their lowest action possible in equilibrium and a different largest Nash equilibrium in which everyone is playing their highest action possible in equilibrium. As shown by results for lattice games in the following chapters, both nonempty Nash equilibrium set and its lattice structure are properties of GSC that do not generalize easily.

Examples 3.21 Consider the following examples.

1. In the coordination game (Example 1.1, page 2), the Nash equilibrium set is $\mathcal{E} = \{(A, A), (B, B)\}$, a nonempty complete lattice. In this case, this is a subcomplete chain in the lattice of profiles of actions. Moreover, each Nash equilibrium is strict and each Nash equilibrium is symmetric.
2. The team trust game (Example 1.4, page 5) has a unique Nash equilibrium, $\mathcal{E} = \{(D, D)\}$.
3. In the multi-player coordination game (Example 1.5, page 7), the Nash equilibrium set has two elements,

$$\mathcal{E} = \{(0, \ldots, 0), (1, \ldots, 1)\}.$$

This is a nonempty complete lattice. Moreover, the equilibrium set is a subcomplete chain. Both equilibria are strict and both equilibria are symmetric.

4. In Bertrand oligopoly (Example 1.7, page 8), there is a unique Nash equilibrium with equilibrium price of each firm i given by $x_i^* = \frac{a-c}{2b-(I-1)\beta}$. It is strict and symmetric. The Nash equilibrium set is trivially a complete lattice.

3.2.4 *Dominance Solvability and Global Stability*

A common theme in monotone games is that monotonicity of best responses implies that undominated responses to an interval are bounded by best response behavior at end points of the interval. In the literature, these bounds are typically derived jointly for all players.

A new and unifying insight here is that it is useful to develop these bounds at the level of an individual player. Joint behavior can then be derived from behavior of different types of individual players in the game. This provides a common thread to understand behavior of different types of players across different classes of monotone games. It also makes the proofs more intuitive, accessible, and uniform. The following notation is helpful.

For each player i and for each $x_{-i}, y_{-i} \in X_{-i}$ with $x_{-i} \preceq y_{-i}$, let $\mathcal{U}_i([x_{-i}, y_{-i}])$ be the set of undominated responses of player i to $[x_{-i}, y_{-i}]$, and let $[\mathcal{U}_i([x_{-i}, y_{-i}])]$ be the interval determined by $\mathcal{U}_i([x_{-i}, y_{-i}])$, that is,

$$[\mathcal{U}_i([x_{-i}, y_{-i}])] = [\inf \mathcal{U}_i([x_{-i}, y_{-i}]), \sup \mathcal{U}_i([x_{-i}, y_{-i}])].$$

For notational convenience, for each $x_{-i} \in X_{-i}$, let $\underline{B}_i(x_{-i}) = \inf B_i(x_{-i})$ and $\overline{B}_i(x) = \sup B_i(x_{-i})$. As $B_i(x_{-i})$ is subcomplete, it follows that $\underline{B}_i(x_{-i})$ is the smallest best response to x_{-i} and $\overline{B}_i(x_{-i})$ is the largest best response to x_{-i}.

Similarly, for profiles of actions $x, y \in X$ with $x \preceq y$, let $\mathcal{U}([x, y])$ denote the set of profiles of undominated responses to $[x, y]$, and let $[\mathcal{U}([x, y])]$ be the interval determined by $\mathcal{U}([x, y])$, that is,

$$[\mathcal{U}([x, y])] = [\inf \mathcal{U}([x, y]), \sup \mathcal{U}([x, y])].$$

For each $x \in X$, let $\underline{B}(x) = \inf B(x)$ and $\overline{B}(x) = \sup B(x)$. As $B(x)$ is subcomplete, it follows that $\underline{B}(x)$ is the profile of smallest best responses to x and $\overline{B}(x)$ is the profile of largest best responses to x.

Theorem 3.22 *In every GSC,*

1. *For every i and for every $x_{-i}, y_{-i} \in X_{-i}$ with $x_{-i} \preceq y_{-i}$,*

$$[\mathcal{U}_i([x_{-i}, y_{-i}])] = [\underline{B}_i(x_{-i}), \overline{B}_i(y_{-i})].$$

2. *For every $x, y \in X$ with $x \preceq y$, $[\mathcal{U}([x, y])] = [\underline{B}(x), \overline{B}(y)]$.*

Proof For statement (1), consider player i and $x_{-i}, y_{-i} \in X_{-i}$ with $x_{-i} \preceq y_{-i}$. In one direction, as both $\underline{B}_i(x_{-i})$ and $\overline{B}_i(y_{-i})$ are best responses, they are both undominated responses to $[x_{-i}, y_{-i}]$, and therefore, both $\underline{B}_i(x_{-i})$ and $\overline{B}_i(y_{-i})$ are in $\mathcal{U}_i([x_{-i}, y_{-i}])$, whence $[\underline{B}_i(x_{-i}), \overline{B}_i(y_{-i})] \subset [\mathcal{U}_i([x_{-i}, y_{-i}])]$.

In the other direction, suppose $\hat{x}_i \in \mathcal{U}_i([x_{-i}, y_{-i}])$ and $\hat{x}_i \notin [\underline{B}_i(x_{-i}), \overline{B}_i(y_{-i})]$. As case 1, suppose $\hat{x}_i \not\preceq \overline{B}_i(y_{-i})$. In this case, $\overline{B}_i(y_{-i}) \vee \hat{x}_i \notin B_i(y_{-i})$. Therefore, for arbitrary $\tilde{x}_{-i} \in [x_{-i}, y_{-i}]$,

$$\begin{aligned} & u_i(\overline{B}_i(y_{-i}) \vee \hat{x}_i, y_{-i}) \not\geq u_i(\overline{B}_i(y_{-i}), y_{-i}) \\ \Rightarrow \quad & u_i(\hat{x}_i, y_{-i}) \not\geq u_i(\overline{B}_i(y_{-i}) \wedge \hat{x}_i, y_{-i}) \\ \Rightarrow \quad & u_i(\hat{x}_i, \tilde{x}_{-i}) \not\geq u_i(\overline{B}_i(y_{-i}) \wedge \hat{x}_i, \tilde{x}_{-i}), \end{aligned}$$

where the first implication follows from u_i being quasisupermodular on X_i, and the second implication follows from u_i having single crossing property in (x_i, x_{-i}). As $\tilde{x}_{-i}$ is arbitrary in $[x_{-i}, y_{-i}]$, it follows that $\overline{B}_i(y_{-i}) \wedge \hat{x}_i$ strictly dominates $\hat{x}_i$, which contradicts $\hat{x}_i \in \mathcal{U}_i([x_{-i}, y_{-i}])$. Similarly, case 2, in which $\underline{B}_i(x_{-i}) \not\preceq \hat{x}_i$ yields a contradiction. Therefore, $[\mathcal{U}_i([x_{-i}, y_{-i}])] \subset [\underline{B}_i(x_{-i}), \overline{B}_i(y_{-i})]$.

For statement (2), consider $x, y \in X$ with $x \preceq y$. Statement (1) shows that for every i, $[\mathcal{U}_i([x_{-i}, y_{-i}])] = [\underline{B}_i(x_{-i}), \overline{B}_i(y_{-i})]$, and therefore, $\inf \mathcal{U}_i([x_{-i}, y_{-i}]) = \underline{B}_i(x_{-i}) \in \mathcal{U}_i([x_{-i}, y_{-i}])$ and $\sup \mathcal{U}_i([x_{-i}, y_{-i}]) = \overline{B}_i(y_{-i}) \in \mathcal{U}_i([x_{-i}, y_{-i}])$. Therefore,

$$\begin{aligned} [\mathcal{U}([x, y])] &= \times_{i=1}^{I} [\mathcal{U}_i([x_{-i}, y_{-i}])] \\ &= \times_{i=1}^{I} [\underline{B}_i(x_{-i}), \overline{B}_i(y_{-i})] \\ &= [\underline{B}(x), \overline{B}(y)], \end{aligned}$$

where the first and third equalities follow from Proposition 2.9 (page 20), and the second equality from statement (1). □

Milgrom and Shannon (1994) prove statement (2) in the previous theorem directly. The proof here isolates and proves the individual player argument separately as statement (1). Statement (2) combines the result from statement (1) for all players in the game by using products of generated intervals given in Theorem 2.9 (page 20). This separation of the two is useful to provide similar granular arguments in the following chapters.

Combined with earlier results on behavior of best response dynamics, the following results due to Milgrom and Shannon (1994) for GSC follow in a straightforward manner. Recall that a sequence $(x^n)_{n=0}^{\infty}$ in X is an adaptive dynamic, if for every $n \geq 0$, there is $k_n > n$, such that for every $k \geq k_n$, $x^k \in [\mathcal{U}[P(n,k)]]$, where $P(n,k) = \{x^m \mid n \leq m < k\}$ is play from n to k (not including k). A lattice game is globally stable if there is $x^* \in X$ such that every adaptive dynamic converges to x^*.

Theorem 3.23 *In a GSC, let* $(\underline{x}^n)_{n=1}^{\infty}$ *and* $(\overline{x}^n)_{n=1}^{\infty}$ *be best response dynamics from* $\inf X$ *and* $\sup X$*, respectively, and* $\underline{x}$ *and* $\overline{x}$ *be their respective limits.*

1. *Every profile of serially undominated strategies is in the interval* $[\underline{x}, \overline{x}]$.
2. $\underline{x}$ *is the smallest profile of serially undominated strategies in the game, and* $\overline{x}$ *is the largest profile of serially undominated strategies in the game.*
3. $\underline{x}$ *is the smallest Nash equilibrium in the game, and* $\overline{x}$ *is the largest Nash equilibrium in the game.*
4. *Every adaptive dynamic* $(x^n)_{n=0}^{\infty}$ *satisfies* $\underline{x} \preceq \liminf x^k \preceq \limsup x^k \preceq \overline{x}$.

Proof For statement (1), Theorem 3.22 (page 68) and definition of best response dynamics imply that for every $n \geq 0$,

$$[\mathcal{U}([\underline{x}^n, \overline{x}^n])] = [\underline{B}(\underline{x}^n), \overline{B}(\overline{x}^n)] = [\underline{x}^{n+1}, \overline{x}^{n+1}],$$

and therefore, the set of serially undominated strategies satisfies

$$\mathcal{U}^{\infty} \subset \bigcap_{n=0}^{\infty} [\mathcal{U}([\underline{x}^n, \overline{x}^n])] = \bigcap_{n=0}^{\infty} [\underline{x}^{n+1}, \overline{x}^{n+1}] = [\underline{x}, \overline{x}].$$

Statement (2) follows, because $\underline{x}$ and $\overline{x}$ are Nash equilibria, and therefore, both are profiles of serially undominated strategies, and combined with statement (1), $\underline{x}$ is the smallest profile of serially undominated strategies in the game, and $\overline{x}$ is the largest profile of serially undominated strategies in the game.

Statement (3) holds, because every Nash equilibrium is serially undominated, and therefore, the set of Nash equilibria is contained in $[\underline{x}, \overline{x}]$, and moreover, Theorem 3.19 (page 65) implies that both $\underline{x}$ and $\overline{x}$ are Nash equilibria.

Statement (4) is proved by first proving that for every $n \geq 0$, there is $k_n > n$, for every $k \geq k_n$, $x^k \in [\underline{x}_n, \overline{x}_n]$. This is proved using induction on n. For $n = 0$, this is trivially true, because $[\underline{x}^0, \overline{x}^0] = X$. Suppose this statement is true for fixed n. Let $k_n > n$ be such that for every $k \geq k_n$, $x^k \in [\underline{x}^n, \overline{x}^n]$. Then for every $k > k_n$, $P(k_n, k) \subset [\underline{x}^n, \overline{x}^n]$. Moreover, applying the definition of adaptive dynamics to index k_n, let $k_{n+1} > k_n$ be such that for every $k \geq k_{n+1}$, $x^k \in [\mathcal{U}([P(k_n, k)])]$. Notice that for every $k \geq k_{n+1}$, $P(k_n, k) \subset [\underline{x}^n, \overline{x}^n]$. Therefore, $k_{n+1} > n + 1$, and for every $k \geq k_{n+1}$,

$$x^k \in [\mathcal{U}([P(k_n, k)])] \subset [\mathcal{U}([\underline{x}^n, \overline{x}^n])] = [\underline{x}^{n+1}, \overline{x}^{n+1}].$$

Statement (4) now follows, because for every $n \geq 0$,

$$\underline{x}^n \preceq \liminf x^k \preceq \limsup x^k \preceq \overline{x}^n,$$

and therefore, $\underline{x} \preceq \liminf x^k \preceq \limsup x^k \preceq \overline{x}$. □

Theorem 3.24 *In every GSC, the following are equivalent.*

1. *The game is globally stable*
2. *The game is dominance solvable*
3. *The game has a unique Nash equilibrium*
4. $\underline{x} = \overline{x}$

Proof The equivalence of statements (2) and (4) follows from statements (1) and (2) of Theorem 3.23 (page 69). The equivalence of statements (3) and (4) follows from statement (3) of Theorem 3.23 (page 69). Statement (4) implies statement (1) follows from statement (4) of Theorem 3.23 (page 69). Statement (1) implies statement (4), because if every adaptive dynamic converges to the same point, then the best response dynamic from $\inf X$ and the best response dynamic from $\sup X$, both of which are adaptive dynamics, converge to the same point, and therefore, $\underline{x} = \overline{x}$. □

An implication is that limiting behavior of every adaptive dynamic in a GSC is bounded by the same extremal Nash equilibria, $\underline{x}$ and $\overline{x}$. Best response dynamics from inf X and sup X provide a constructive method to compute these bounds. In particular, with finitely many actions, these bounds are attained in finitely many iterations. Moreover, when the game has a unique Nash equilibrium, it is globally stable.

As mentioned in Sect. 2.2.4 (page 38), serially undominated strategies implicitly assume thoughtful players, thinking about other thoughtful players, using infinite iterations of introspection about reasonable actions to arrive at possible solutions to the game. The uniqueness of such a solution is a robust prediction for this type of behavior.

On the other hand, adaptive dynamics implicitly assume myopic players, adapting dynamically over time to past play, and using convergence of such behavior to arrive at a possible solution to the game. When all adaptive behavior converges to the same limit, that limiting behavior is a robust prediction for outcome in the game.

In GSC and in other classes of monotone games in the following chapters, these two different foundations for robustness of predicted outcomes are equivalent.

GSC has the special additional feature that each of these properties is further equivalent to uniqueness of Nash equilibrium. This is not true more generally. Therefore, a useful way to check for robustness of outcomes in GSC is to inquire whether the Nash equilibrium set in such a game is a singleton or not. This may be achieved computationally by comparing the limit of best response dynamic from inf X to limit of best response dynamic from sup X, or equivalently, by testing that the sequence derived by taking the difference of these two best response dynamics converges to zero.

Examples 3.25 Consider the following examples.

1. The coordination game (Example 1.1, page 2) with $A \prec B$ has two Nash equilibria, $\mathcal{E} = \{(A, A), (B, B)\}$. The interval $[(A, A), (B, B)]$ is the whole space of profiles of actions. Therefore, bounds on serially undominated strategies do not restrict any profile and the game is not globally stable.
2. The team trust game (Example 1.4, page 5) has a unique Nash equilibrium, $\mathcal{E} = \{(D, D)\}$. Therefore, the game is dominance solvable and globally stable.

3. The multi-player coordination game (Example 1.5, page 7) has two Nash equilibria, the smallest profile of actions and the largest one.

$$\mathcal{E} = \{(0, \dots, 0), (1, \dots, 1)\}.$$

The interval formed by these is the whole space of profiles. Therefore, bounds on serially undominated strategies do not restrict any profile and the game is not globally stable.
4. Bertrand oligopoly (Example 1.7, page 8), has a unique Nash equilibrium with equilibrium price of each firm i given by $x_i^* = \frac{a-c}{2b-(I-1)\beta}$. Therefore, the game is dominance solvable and globally stable.

Example 3.26 (Discrete Bertrand duopoly) Consider two Indian restaurants competing as Bertrand duopolists. Each restaurant can choose to sell lower quality restaurant meal for lower prices or higher quality restaurant meal for higher prices. There are four quality-price categories ranked from low (L) to medium-low (ML) to medium-high (MH) to high (H), with $L \prec ML \prec MH \prec H$.

Suppose market conditions are as follows. If firm 2 prices low or medium-low, it is in firm 1's best interest to price low. It can survive if it prices medium-low, but cannot survive at higher prices, because it loses too many customers. If firm 2 prices medium-high or high, it is in firm 1's best interest to price medium-high. It can survive if it prices high, but cannot survive at lower prices, because its output is considered very low quality and it loses too many customers. Conditions for firm 2 are motivated similarly. An assignment of payoffs with these features is given in the bimatrix in Table 3.2 (page 72). This is a GSC. The best response is (weakly) increasing and singleton valued and Nash equilibrium set is $\{(L, L), (MH, MH)\}$. This is a complete lattice and every equilibrium is symmetric.

Table 3.2 Discrete Bertrand duopoly

		F2			
		L	ML	MH	H
F1	L	2, 2	2, 1	0, 0	0, 0
	ML	1, 2	1, 1	0, 0	0, 0
	MH	0, 0	0, 0	2, 2	2, 1
	H	0, 0	0, 0	1, 2	1, 1

Serially undominated strategies lie in the interval $[(L, L), (MH, MH)]$, which is smaller than the whole space of profiles of actions. This game is neither dominance solvable nor globally stable.

3.2.5 Monotone Comparative Statics of Equilibrium

For situations with codirectional incentives, the effect of a change in the decision-making environment on predicted outcomes is studied using parameterized games with strategic complements. These are a subclass of parameterized lattice games.

In a parameterized lattice game $\Gamma = ((X_i, u_i)_{i=1}^{I}, T)$, for each parameter $t \in T$, ***player i has strategic complements at t***, if the section of player i payoff determined by t, $u_i(x_i, x_{-i}, t)$, has single crossing property in (x_i, x_{-i}) and is quasisupermodular on X_i (for each x_{-i}).

As motivated in the section on parameterized lattice games, in situations where changes in the decision-making environment are important, the question of interest may be formulated in the following manner. If decision-making environment is made more favorable to take a higher action, when will equilibrium actions go up? As shown by the parameterized discrete Bertrand duopoly (Example 3.28, page 75), this is not always the case.

The idea that an increase in parameter t provides to player i an incentive to take a higher action is formalized as follows. ***Player i has incentive to take higher action in t***, if $u_i(x_i, x_{-i}, t)$ has single crossing property in (x_i, t), for every x_{-i}.

A ***parameterized game with strategic complements***, or ***parameterized GSC***, is a parameterized lattice game $\Gamma = ((X_i, u_i)_{i=1}^{I}, T)$, where

1. There are finitely many players, I, indexed by $i \in \{1, \ldots, I\}$.
2. For each i, the action space of player i is X_i. It is assumed to be a nonempty, subcomplete lattice in $\mathbb{R}^{n_i}$ with the Euclidean order. The space of profiles of actions is $X = X_1 \times \cdots \times X_I$ with the product order. The parameter space is $T \subset \mathbb{R}^n$. It is a poset with the Euclidean partial order.
3. For each i, the payoff for player i is $u_i : X_i \times X_{-i} \times T \to \mathbb{R}$, where

   (a) u_i is upper semicontinuous on X and continuous on X_{-i}, for every t,

   (b) u_i has single crossing property in (x_i, x_{-i}) (for each t), u_i is quasisupermodular on X_i (for each (x_{-i}, t)), and

(c) u_i has single crossing property in (x_i, t) (for each x_{-i}).

In other words, a parameterized GSC is a parameterized lattice game in which every player i has strategic complements at every t and every player i has incentive to take a higher action in t.

It follows immediately from the definition that in a parameterized GSC, for every $t \in T$, the game at t, $\Gamma(t)$, is a GSC. In particular, the section of best response correspondence for player i determined by t, $B_i(\cdot, t)$, mapping (x_{-i}, t) to $B_i(x_{-i}, t)$ is nonempty valued, compact valued, subcomplete valued, upper hemicontinuous, and (weakly) increasing.

Moreover, single crossing property in (x_i, t) implies that for each x_{-i}, the section of best response correspondence for player i determined by x_{-i}, $B_i(x_{-i}, \cdot)$ mapping (x_{-i}, t) to $B_i(x_{-i}, t)$ is (weakly) increasing.

Similarly, for each t, the section of joint best response correspondence determined by t, $B(\cdot, t)$, mapping (x, t) to $B(x, t)$ is nonempty valued, compact valued, subcomplete valued, upper hemicontinuous, and (weakly) increasing, and moreover, for each x, the section of joint best response correspondence determined by x, $B(x, \cdot)$ mapping (x, t) to $B(x, t)$ is (weakly) increasing.

Furthermore, $\Gamma(t)$ is a GSC implies that the equilibrium correspondence $\mathcal{E} : T \rightrightarrows X$, mapping t to $\mathcal{E}(t)$ is nonempty valued and complete lattice valued. For each $t \in T$, let $\underline{e}(t) = \inf_{\mathcal{E}(t)} \mathcal{E}(t)$ be the smallest Nash equilibrium at t and $\overline{e}(t) = \sup_{\mathcal{E}(t)} \mathcal{E}(t)$ be the largest Nash equilibrium at t.

In parameterized GSS, both $\underline{e}(t)$ and $\overline{e}(t)$ are (weakly) increasing functions of t, as shown in Milgrom and Shannon (1994). Existence of increasing equilibrium selections is due to Lippman et al. (1987).

Theorem 3.27 (MCS-GSC) *In every parameterized GSC, for every $\hat{t}, \tilde{t} \in T$, if $\hat{t} \preceq \tilde{t}$, then $\underline{e}(\hat{t}) \preceq \underline{e}(\tilde{t})$ and $\overline{e}(\hat{t}) \preceq \overline{e}(\tilde{t})$.*

Proof Consider $\hat{t} \preceq \tilde{t}$ and let $x^* = \overline{e}(\hat{t})$ and $y^* = \overline{e}(\tilde{t})$. By Theorem 2.18 (page 25),

$$x^* = \sup_X \{x \in X \mid B(x, \hat{t}) \cap [x, \sup X] \neq \emptyset\}.$$

As x^* is a Nash equilibrium at $\hat{t}$, $x^* \in B(x^*, \hat{t})$. Combined with $B(x^*, \hat{t}) \sqsubseteq B(x^*, \tilde{t})$, it follows that $x^* \preceq \sup_X B(x^*, \tilde{t})$. Moreover, $B(x^*, \tilde{t})$ is subcomplete implies that $\sup_X B(x^*, \tilde{t}) \in B(x^*, \tilde{t})$. Therefore, $\sup_X B(x^*, \tilde{t}) \in B(x^*, \tilde{t}) \cap [x^*, \sup X]$, showing that $B(x, \tilde{t}) \cap [x, \sup X]$ is not empty, and consequently, x^* is in the set $\{x \in X \mid B(x, \tilde{t}) \cap [x, \sup X] \neq \emptyset\}$. As

$$y^* = \sup_X \{x \in X \mid B(x, \tilde{t}) \cap [x, \sup X] \neq \emptyset\},$$

it follows that $x^* \preceq y^*$, as desired. The proof that $\underline{e}(\hat{t}) \preceq \underline{e}(\tilde{t})$ is similar. □

In other words, when parameter increase provides each player an incentive to take a higher action, if the parameter goes up, then the action of every player goes up (weakly) in the smallest equilibrium before and after the change, and the action of every player goes up (weakly) in the largest equilibrium before and after the change. This is termed ***monotone comparative statics of extremal equilibrium outcomes.*** A simple example is given by the parameterized coordination game in Example 2.32 (page 43).

As equilibrium profiles $\underline{e}(t)$ and $\bar{e}(t)$ are the smallest and largest serially undominated strategies in the GSC at t, respectively, these may be computed algorithmically using the lower and upper best response dynamics, respectively.

If the game has a unique equilibrium before and after the parameter change, Theorem 3.27 (page 74) implies that this equilibrium profile goes up (weakly). More generally, this theorem does not necessarily imply that the action of a player in every equilibrium at a higher parameter is higher than their action in every equilibrium at a lower parameter.

Example 3.28 (Discrete Bertrand duopoly, part 2) Consider the discrete Bertrand duopoly (Example 3.26, page 72).

Now suppose demand conditions improve, perhaps because the economy is doing well, or more people like Indian food. New market conditions are as follows. If firm 2 prices low or medium-low, it is in firm 1's best interest to price medium-low. It can survive if it prices low, but cannot survive at higher prices, because it loses too many customers. If firm 2 prices medium-high or high, it is in firm 1's best interest to price high. It can survive if it prices medium-high or medium-low, but cannot survive at low price, because its output is considered very low quality and it loses too many customers. Conditions for firm 2 are motivated similarly. An assignment of payoffs with these features is given in the bimatrix in panel (2) of Table 3.3 (page 76). Panel (1) reproduces the payoffs from Table 3.2 (page 72). The new game remains a GSC. The best response is (weakly) increasing and singleton valued and Nash equilibrium set is $\{(ML, ML), (H, H)\}$. This is a complete lattice and every equilibrium is symmetric.

Table 3.3 Discrete Bertrand duopoly, part 2

(1)

F1 \ F2	L	ML	MH	H
L	2, 2	2, 1	0, 0	0, 0
ML	1, 2	1, 1	0, 0	0, 0
MH	0, 0	0, 0	2, 2	2, 1
H	0, 0	0, 0	1, 2	1, 1

(2)

F1 \ F2	L	ML	MH	H
L	1, 1	1, 2	0, 0	0, 0
ML	2, 1	2, 2	1, 0	1, 0
MH	0, 0	0, 1	2, 2	2, 3
H	0, 0	0, 1	3, 2	3, 3

Suppose initial demand (or economic) conditions are summarized by parameter $t = 0$ and new (better or higher) conditions by $t = 1$. Nash equilibrium set at $t = 0$ is $\mathcal{E}(0) = \{(L, L), (MH, MH)\}$ and at $t = 1$ is $\mathcal{E}(1) = \{(ML, ML), (H, H)\}$. When the demand (or economic) conditions improve, the smallest equilibrium moves up from (L, L) to (ML, ML) and the largest equilibrium moves up from (MH, MH) to (H, H).

An increase in the smallest equilibrium and an increase in the largest equilibrium does not necessarily imply that every equilibrium at the lower parameter is lower than every equilibrium at the higher parameter. In this example, in the equilibrium (MH, MH) at the lower parameter value, both firms take higher actions as compared to their actions in the equilibrium (ML, ML) at the higher parameter value.

Examples 3.29 (Parameterized Bertrand oligopoly) In Bertrand oligopoly (Example 1.7, page 8), profit of firm i from setting price x_i when opponents set prices x_{-i} is given by $u_i(x_i, x_{-i}) = (x_i - c)(a - bx_i + \beta(\sum_{j \neq i} x_j))$, best response of firm i is $B_i(x_{-i}) = \frac{1}{2b}(a - c + \beta(\sum_{j \neq i} x_j))$, and the unique (and symmetric) Nash equilibrium price is $x_i^* = \frac{a-c}{2b-(I-1)\beta}$.

We may view a, b and β as demand parameters and c as a cost parameter summarizing the decision-making environment for the firms. If consumer preferences or consumer wealth changes, it may lead to an increase in a and an increase in equilibrium price. An increase in competitor substitutability parameter β (perhaps due to negative ratings received by competitors or increase in reputation of firm i product) also increases the price. An improvement in manufacturing technology may lead to a decrease in c (or an increase in $-c$) and an increase in equilibrium price. More generally, when the multidimensional parameter $(a, -b, \beta, -c)$ goes up, the equilibrium price goes up as well.

This chapter presents foundations of games with strategic complements. Work in this area continues to grow in several directions, including auctions, global games, mechanism design, networks, optimization theory, and decision-making under uncertainty. The next chapter focuses on games with strategic substitutes.

References

Acemoglu, D., & Jensen, M. K. (2013). Aggregate comparative statics. *Games and Economic Behavior*, *81*, 27–49.

Amir, R., Jakubczyk, M., & Knauff, M. (2008). Symmetric versus asymmetric equilibria in symmetric supermodular games. *International Journal of Game Theory*, *37*, 307–320.

Bulow, J. I., Geanakoplos, J. D., & Klemperer, P. D. (1985). Multimarket Oligopoly: Strategic Substitutes and Complements. *Journal of Political Economy*, *93*(3), 488–511.

Dixit, A. (1979). A Model of Duopoly Suggesting a Theory of Entry Barriers. *Bell Journal of Economics*, *10*, 20–32.

Dixit, A. (1980). The Role of Investment in Entry-Deterrence. *Economic Journal*, *90*, 95–106.

Echenique, F. (2002). Comparative statics by adaptive dynamics and the correspondence principle. *Econometrica*, *70*(2), 257–289.

Echenique, F., & Sabarwal, T. (2003). Strong Comparative Statics of Equilibria. *Games and Economic Behavior*, *42*(2), 307–314.

Fudenberg, D., & Tirole, J. (1984). The fat-cat effect, the puppy-dog ploy, and the lean and hungry look. *American Economic Review*, *74*(2), 361–366.

Jensen, M. K. (2010). Aggregative Games and Best-Reply Potentials. *Economic Theory*, *43*(1), 45–66.

Lippman, S. A., Mamer, J. W., & McCardle, K. F. (1987). Comparative Statics in non-cooperative games via transfinitely iterated play. *Journal of Economic Theory*, *41*(2), 288–303.

Milgrom, P., & Roberts, J. (1990). Rationalizability, learning, and equilibrium in games with strategic complementarities. *Econometrica*, *58*(6), 1255–1277.

Milgrom, P., & Shannon, C. (1994). Monotone Comparative Statics. *Econometrica*, *62*(1), 157–180.

Quah, J. K.-H., & Strulovici, B. (2009). Comparative statics, informativeness, and the interval dominance order. *Econometrica*, *77*(6), 1949–1992.

Shannon, C. (1995). Weak and Strong Monotone Comparative Statics. *Economic Theory*, *5*(2), 209–227.

Sobel, M. (1988). *Isotone comparative statics in supermodular games*. SUNY at Stony Brook: Mimeo.

Spence, A. M. (1977). Entry, Capacity, Investment and Oligopolistic Pricing. *Bell Journal of Economics*, *8*, 534–44.

Topkis, D. (1978). Minimizing a submodular function on a lattice. *Operations Research*, *26*, 305–321.

Topkis, D. (1979). Equilibrium points in nonzero-sum n-person submodular games. *SIAM Journal on Control and Optimization*, *17*(6), 773–787.

Topkis, D. (1998). *Supermodularity and Complementarity*. Princeton University Press.

Vives, X. (1990). Nash Equilibrium with Strategic Complementarities. *Journal of Mathematical Economics*, *19*(3), 305–321.

Vives, X. (1999). *Oligopoly pricing*. Cambridge, MA and London, UK: MIT Press.

Zhou, L. (1994). The Set of Nash Equilibria of a Supermodular Game is a Complete Lattice. *Games and Economic Behavior*, *7*(2), 295–300.

CHAPTER 4

Games with Strategic Substitutes

Abstract Contradirectional incentives formalize situations in which participants have an incentive to move in the direction opposite to other participants. Payoff functions with a combination of decreasing differences, submodularity, dual single crossing property, and quasisubmodularity capture these incentives. Games with strategic substitutes are lattice games in which each player has an incentive to move in the direction opposite to their opponents. Distinctive features of these games imply new patterns for individual and equilibrium behavior. The theory identifies principles governing behavior in these cases and a diversity of examples highlights applications of these principles.

Keywords Contradirectional incentives · Strategic substitutes · Decreasing maximizers · Opposing direct and indirect effects · Totally unordered equilibrium set · Never decreasing equilibrium

Although both strategic complements and strategic substitutes arise frequently in many socioeconomic situations, the focus in the earlier literature was on the case of strategic complements.

As shown in Chapter 3 (page 45), games with strategic complements have several appealing characteristics. They always have a Nash equilibrium and the Nash equilibrium set is a nonempty complete lattice. Moreover, when environmental parameters provide players an incentive to take higher

T. Sabarwal, *Monotone Games*,
https://doi.org/10.1007/978-3-030-45513-2_4

action, the action of every player in the extremal equilbria goes up with the parameter providing an intuitive result for monotone comparative statics of extremal equilibrium outcomes.

For environments beyond strategic complements, a standard approach in the earlier literature was to try and modify those environments to fit the same mold as strategic complements by changing the partial order. This change in partial order worked intuitively in special cases only. The more general transformed environments may bear less resemblance to their original motivation and interact in counter intuitive ways in parameterized games, further limiting their usefulness. Direct results for more general cases were unavailable.

More recent work has shown explicitly that more general environments may not necessarily be transformed into strategic complements or be amenable to the same techniques. (See Roy and Sabarwal (2008), Roy and Sabarwal (2010), Roy and Sabarwal (2012), and Monaco and Sabarwal (2016).)

This work shows that a game with strategic substitutes (GSS) may not necessarily have a Nash equilibrium, in contrast to the situation for GSC, in which the Nash equilibrium set is always nonempty. Therefore, in general, it is impossible to change a GSS into a GSC by changing the partial order.

Moreover, with strategic substitutes for one or two players only, the Nash equilibrium set in a lattice game is totally unordered. Therefore, with multiple equilibria, it is impossible to talk about extremal equilibria (because they would be ordered), and there is no hope to have monotone comparative statics of extremal equilibrium outcomes.

These results show that attempts to fit strategic substitutes into the mold of strategic complements may be problematic. New tools are needed to study these cases directly.

This chapter develops the theory of games with strategic substitutes including results for more general lattice games. Distinctive features of these games imply new patterns for individual and equilibrium behavior.

The general theory of contradirectional incentives is developed and used to formulate conditions under which best responses are decreasing in opponent actions. This is used to define when a player has strategic substitutes.

An example shows that the Nash equilibrium set in a GSS may be empty. This is shown for a three player, two action GSS, the simplest possible case beyond a two player game.

In a general lattice game, with strategic substitutes for one or two players only, the Nash equilibrium set is totally unordered. No additional conditions are needed on payoffs of other players. Other players may have strategic substitutes, or strategic complements, or neither, and the Nash equilibrium set remains totally unordered. As a special case, in a GSS in which these conditions are satisfied, the Nash equilibrium set is totally unordered. As another special case, starting with a GSC and adding one player with a strictly decreasing best response function makes the Nash equilibrium set totally unordered.

This shows that the lattice structure of Nash equilibrium set in a GSC is not robust to minimal deviations from GSC while the totally unordered Nash equilibrium set in a GSS generalizes to other lattice games. These results also provide mild conditions under which a lattice game can have at most one symmetric equilibrium. This is in contrast to GSC as well.

With strategic substitutes, the incentive of a player to move opposite to their opponents implies differences in best response dynamics as compared to the case of strategic complements, where the incentive of a player is to move together with other players. New versions of best response dynamics are needed to study strategic substitutes and their limiting behavior is different in important ways from that of strategic complements. In particular, convergence of a single best response dynamic (either from $\inf X$ or from $\sup X$) is equivalent to dominance solvability and global stability, in contrast to the result for GSC. In general, uniqueness of Nash equilibrium is not a sufficient condition for this result.

Change in equilibrium behavior in a parameterized GSS is different from that of parameterized GSC in important ways and this difference generalizes to parameterized lattice games. When each player has an incentive to take higher action in the parameter, under mild conditions on one or two players only, the equilibrium correspondence in a parameterized lattice game is never decreasing. That is, for every equilibrium at a low parameter value and every equilibrium at a higher parameter value, the new equilibrium cannot be lower than the old equilibrium. No additional conditions are needed on payoffs for other players. As a special case, in a parameterized GSS in which these conditions are satisfied, the equilibrium correspondence is never decreasing. This is in contrast to the result for parameterized GSC. As another special case, this implies that starting with a parameterized GSC and adding one player with strictly decreasing best response function, all decreasing selections of equilibria are ruled out.

These results provide similarly mild conditions under which a parameterized lattice game has a unique symmetric equilibrium that is increasing in the parameter. This monotone comparative statics for symmetric Nash equilibrium is not true in general for a parameterized GSC. In that case, there may be multiple symmetric Nash equilibria and a symmetric Nash equilibrium at a lower parameter value is not necessarily lower than a symmetric Nash equilibrium at a higher parameter value. In other words, the result for parameterized GSS generalizes to parameterized lattice games whereas the result for parameterized GSC does not.

For a general parameterized GSS, monotone comparative statics of equilibrium outcomes is obtained by identifying explicitly the impact of opposing effects arising from a parameter increase.

The direct effect of a parameter increase is to provide an incentive for each player to take a higher action, but when opponents take higher actions, the indirect strategic substitutes effect is to take a lower action. An example shows that the indirect effect may dominate the direct effect and the new equilibrium may not have higher actions for all players even if the equilibrium is unique.

A critical insight is that when a composition of these effects is favorable, monotone comparative statics of equilibrium outcomes reemerges. The composite effect blends the right intuition about direct and indirect effects and yields an appropriately nuanced version of monotone comparative statics of equilibrium outcomes that holds for more general environments. Additional analysis provides conditions that make it easier to evaluate these competing tradeoffs and provides practical guidelines to apply the result.

These results show that several properties of GSC do not survive with minimal extensions to include strategic substitutes. On the other hand, several properties for GSS extend to include players with strategic complements and to more general lattice games. The overall picture that emerges is that properties of GSC are more special in nature.

4.1 Contradirectional Incentives

Strategic substitutes formalize a class of interdependent decisions that exhibit patterns of opposition, or movement in the direction opposite to other decision-makers. As mentioned in Chapter 1 (page 1), this basic principle emerges in many different socioeconomic environments involving competition for a shared resource such as market for a good, a public park, police services, river water use, roads and transportation, public education, limited funds for research grants, and so on.

In each instance, the more intensively others use the shared resource, the best response for a given participant is to use it less intensively, because their marginal benefit (or probability of success) from more intensive use decreases when others use the resource more intensively.

One way to think about incentive for a player to take a lower (or opposite) action when opponents take a higher action is that marginal benefit to the player from a lower action goes up when opponents take a higher action. This intuition is formalized by the property of decreasing differences. The notion of decreasing differences generalizes to that of submodular function on a lattice.

Decreasing differences and submodularity are both cardinal properties that do not necessarily hold under monotonic transformations of a payoff function. The more general notions of dual single crossing property and quasisubmodular function provide this additional flexibility. These properties provide a mathematical foundation for contradirectional incentives.

Dual single crossing property and quasisupermodularity are used to characterize when maximizers in an optimization problem are decreasing in the parameters of the problem. This provides a mathematical foundation for strategic substitutes and for statements of the form: If actions of opponents go up, then best choice of a given player goes down.

4.1.1 Decreasing Differences

For posets $(X, \preceq_X)$ and $(Y, \preceq_Y)$, a function $f : X \times Y \to \mathbb{R}$ has ***decreasing differences in*** (x, y), if for every $\hat{x} \preceq_X \tilde{x}$ and for every $\hat{y} \preceq_Y \tilde{y}$, $f(\tilde{x}, \hat{y}) - f(\hat{x}, \hat{y}) \geq f(\tilde{x}, \tilde{y}) - f(\hat{x}, \tilde{y})$. In other words, for every $\hat{x} \preceq_X \tilde{x}$, the difference $f(\tilde{x}, y) - f(\hat{x}, y)$ is (weakly) decreasing in y. Rearranging terms shows that the order of x and y in the definition does not matter. The following proposition connects decreasing differences and increasing differences.

Proposition 4.1 *Let* $(X, \preceq_X), (Y, \preceq_Y)$ *be posets and* $f : X \times Y \to \mathbb{R}$. f *has decreasing differences in* (x, y), *if, and only if,* $-f$ *has increasing differences in* (x, y).

Proof Let $g = -f$ and notice that for every $\hat{x} \preceq_X \tilde{x}$ and for every $\hat{y} \preceq_Y \tilde{y}$,

$$\begin{aligned} & g(\tilde{x}, \hat{y}) - g(\hat{x}, \hat{y}) \leq g(\tilde{x}, \tilde{y}) - g(\hat{x}, \tilde{y}) \\ \Leftrightarrow\ & f(\hat{x}, \hat{y}) - f(\tilde{x}, \hat{y}) \leq f(\hat{x}, \tilde{y}) - f(\tilde{x}, \tilde{y}) \\ \Leftrightarrow\ & f(\tilde{x}, \hat{y}) - f(\hat{x}, \hat{y}) \geq f(\tilde{x}, \tilde{y}) - f(\hat{x}, \tilde{y}), \end{aligned}$$

and from this, the statement follows immediately. □

In a lattice game $(X_i, u_i)_{i=1}^I$, if payoff of player i has decreasing differences in (x_i, x_{-i}), the marginal benefit to player i from taking a higher action over a lower action, $u_i(\tilde{x}_i, x_{-i}) - u_i(\hat{x}_i, x_{-i})$, is a decreasing function of opponent actions. Equivalently, the marginal benefit to player i from taking a lower action over a higher action, $u_i(\hat{x}, y) - u_i(\tilde{x}, y)$, is (weakly) increasing in y. This provides player i an incentive to take a lower action when opponents take higher actions. In a lattice game, the notion that player i will take lower actions when their opponents take higher actions is sometimes formalized by requiring that payoff of player i has decreasing differences in (x_i, x_{-i}).

For a function on a product space, decreasing differences in particular components of the product space are formalized using the same notation as for increasing differences. Consider posets $X_1, \ldots, X_N$ and let $X = X_1 \times \cdots \times X_N$. A function $f : X \to \mathbb{R}$ has ***decreasing differences in*** (x_i, x_j), for $i \neq j$, if for every $x_{-(i,j)} \in X_{-(i,j)}$, $f(x_i, x_j, x_{-(i,j)})$ has decreasing differences in (x_i, x_j). A function $f : X \to \mathbb{R}$ has ***decreasing differences on*** X, if for every $i \neq j$, f has decreasing differences in (x_i, x_j).

In a lattice game, $(X_i, u_i)_{i=1}^I$, the notion that player i tends to take lower actions when their opponents take higher actions is sometimes formalized by requiring that payoff of player i has decreasing differences in (x_i, x_{-i}).

Examples 4.2 Consider the following examples.

1. In dove hawk game (Example 1.2, page 3), the payoff of each player has decreasing differences. If $D \prec H$, then $u_1(H, D) - u_1(D, D) = 1 \geq -1 = u_1(H, H) - u_1(D, H)$. If $H \prec D$, then $u_1(D, H) - u_1(H, H) = 1 \geq -1 = u_1(D, D) - u_1(H, D)$. Therefore, for each fixed partial order on $\{D, H\}$ in which D and H are comparable, player 1 payoff has decreasing differences. Similarly, player 2 payoff has decreasing differences.
2. In matching pennies (Example 1.3, page 4), for each fixed partial order on $\{H, T\}$ in which H and T are comparable, player 1 payoff has decreasing differences, and player 2 payoff does not. If $H \prec T$, then $u_1(T, H) - u_1(H, H) = 2 \geq -2 = u_1(T, T) - u_1(H, T)$, and $u_2(T, H) - u_2(H, H) = -2 \not\geq 2 = u_2(T, T) - u_2(H, T)$. If $T \prec H$, then $u_1(H, T) - u_1(T, T) = 2 \geq -2 = u_1(H, H) - u_1(T, H)$, and $u_2(H, T) - u_2(T, T) = -2 \not\geq 2 = u_2(H, H) - u_2(T, H)$.

3. In the team trust game (Example 1.4, page 5), payoff differences are constant for both players, and therefore, payoff for each player satisfies definition of decreasing differences.
4. In the multi-player externality game (Example 1.6, page 7), each player i has decreasing differences, because when x_{-i} increases from the zero vector 0 to a nonzero vector $x_{-i}, u_i(1, 0) - u_i(0, 0) = 1 - c \geq 1 - c - 1 = u_i(1, x_{-i}) - u_i(0, x_{-i})$. (The difference $u_i(1, x_{-i}) - u_i(0, x_{-i})$ is constant at $-c$ when evaluated at two different nonzero x_{-i}.)

4.1.2 *Submodular Function*

Another way in which incentives to take lower actions are formalized is by submodular functions, which encapsulate decreasing differences.

Let $(X, \preceq)$ be a lattice. A function $f : X \to \mathbb{R}$ is ***submodular on*** X, if for every $x, y \in X$, $f(x) - f(x \wedge y) \geq f(x \vee y) - f(y)$. If x and y are comparable, the condition holds trivially, and therefore, a nontrivial application requires x and y to be noncomparable. Moreover, if X is a chain, then every function $f : X \to \mathbb{R}$ is submodular. In particular, every function $f : \mathbb{R} \to \mathbb{R}$ is submodular. Furthermore, it follows immediately from the definition that f is submodular on X, if, and only if, $-f$ is supermodular on X.

For a function on a product space, submodular property in a component of the product space is formalized using the same notation as for supermodular property on a component space. Consider lattices $X_1, \ldots, X_N$ and let $X = X_1 \times \cdots \times X_N$ be the product lattice. For $i = 1, \ldots, N$, a function $f : X \to \mathbb{R}$ is ***submodular on*** X_i, if for every $x_{-i} \in X_{-i}$, $f(\cdot, x_{-i})$ is submodular on X_i.

Theorem 4.3 *Let $X_1, \ldots, X_N$ be lattices, $X = X_1 \times \cdots \times X_N$ be the product lattice, and $f : X \to \mathbb{R}$.*

1. *If f is submodular on X, then f has decreasing differences on X.*
2. *If f has decreasing differences on X, and for every i, f is submodular on X_i, then f is submodular on X.*
3. *Suppose $X_1, \ldots, X_N$ are chains.*
   f is submodular on X, if, and only if, f has decreasing differences on X.

Proof Recall that f has decreasing differences on X, if, and only if, $-f$ has increasing differences on X, and f is submodular on X, if, and only if, $-f$ is supermodular on X. Statements (1) and (2) follow from these facts and an application of the results in Theorem 3.2 (page 50) to $-f$. For statement (3), sufficiency follows from statement (1) and necessity follows from statement (2), because every function on a chain is submodular, and therefore, for every i, f is submodular on X_i. □

In statement (2), the condition that f is submodular on X_i cannot be dropped in general.

Example 4.4 Let $X_1 = \{a, b, c, d\}$ with $a \prec b \prec d, a \prec c \prec d$, and b and c are not comparable. Let $X_2 = \{0, 1\}$ with $0 \prec 1$ and consider $X = X_1 \times X_2$ with the product order. Let $f : X \to \mathbb{R}$ be given by

$$f(a, 0) = 5,\ f(b, 0) = 2,\ f(c, 0) = 4,\ f(d, 0) = 3,$$
$$f(a, 1) = 5,\ f(b, 1) = 1,\ f(c, 1) = 3,\ f(d, 1) = 0.$$

Then f has decreasing differences in (x_1, x_2) and in (x_2, x_1), but f is not submodular on X, because if we let $y = (b, 0)$ and $z = (c, 0)$, then $f(y) - f(y \wedge z) = -3 \ngeq -1 = f(y \vee z) - f(z)$. The conclusion does not hold, because $f(\cdot, 0)$ is not submodular on X_1. Also, $f(\cdot, 1)$ is not submodular on X_1.

Theorem 4.5 *Let X, Y be open sublattices of $\mathbb{R}$ and $f : X \times Y \to \mathbb{R}$ be twice continuously differentiable.*
f has decreasing differences on $X \times Y$, if, and only if, $\frac{\partial^2 f}{\partial y \partial x} \leq 0$ on $X \times Y$.

Proof This is proved by applying the results of Theorem 3.4 (page 52) to $-f$ and noting that f has decreasing differences on $X \times Y$, if, and only if, $-f$ has increasing differences on $X \times Y$, and that $\frac{\partial^2 f}{\partial y \partial x} = -\frac{\partial^2 (-f)}{\partial y \partial x}$. □

Combining this with the earlier result yields the following theorem.

Theorem 4.6 *Let $X = X_1 \times \cdots \times X_N$ be an open sublattice of $\mathbb{R}^N$ and $f : X \to \mathbb{R}$ is twice continuously differentiable. The following are equivalent.*

1. *f is submodular on X.*
2. *f has decreasing differences on X.*

3. *For every* $i, j = 1, \ldots, N, i \neq j, \frac{\partial^2 f}{\partial x_i \partial x_j} \leq 0$ *on* $X_i \times X_j$.

Proof The equivalence of (1) and (2) is given in statement (3) of Theorem 4.3 (page 85). The equivalence of (2) and (3) follows from Theorem 4.5 (page 86). □

This characterization provides a convenient tool to check when a function is submodular or satisfies decreasing differences.

Examples 4.7 Consider the following examples.

1. The Cobb–Douglas and Leontief functions in Examples 3.6 (page 53) translate into examples of submodular functions by taking the negative of each function.
2. For lattice games, it is useful to rewrite the following class of examples using contradirectional incentives. In a two player lattice game (X_1, u_1, X_2, u_2), a useful construction to include contradirectional incentives is to think of payoff as a difference of personal effect and interactive effect. For example, suppose $u_1(x_1, x_2) = f(x_1) - \phi(x_1, x_2)$ where f captures personal effect of player 1 action and $-\phi$ captures interactive effect of actions of both players on player 1 payoff. It is immediate that u_1 has decreasing differences in (x_1, x_2), if, and only if, $\phi(x_1, x_2)$ has increasing differences in (x_1, x_2). One way in which this is achieved is when ϕ has a multiplicatively separable form, that is, $\phi(x_1, x_2) = g(x_1)h(x_2)$, where both g and h are (weakly) increasing functions. In particular, if X_1 and X_2 are open sublattices of $\mathbb{R}$ and u_1 is twice continuously differentiable, then

$$\frac{\partial^2 u_1}{\partial x_2 \partial x_1} = -g'(x_1)h'(x_2) \leq 0.$$

Examples of this formulation include the following common forms useful in applications: $\phi(x_1, x_2) = x_1 x_2$, $\phi(x_1, x_2) = \sqrt{x_1 x_2}$, $\phi(x_1, x_2) = x_1\sqrt{x_2}$, $\phi(x_1, x_2) = x_1 \ln x_2$, $\phi(x_1, x_2) = -x_1 e^{-x_2}$, and $\phi(x_1, x_2) = A x_1^{\alpha} x_2^{\beta}$ for $A, \alpha, \beta > 0$, and where $x_1, x_2 \geq 0$, as needed. These generalize to additional cases and to more than two players in a natural manner.

3. In Cournot oligopoly (Example 1.8, page 9), profit of firm i from producing quantity x_i when opponents produce quantities x_{-i} is given by

$$u_i(x_i, x_{-i}) = \left(a - b\left(\sum_{i=1}^{I} x_i\right)\right) x_i - cx_i.$$

For every firm i and for every competitor $j \neq i$, $\frac{\partial^2 u_i}{\partial x_j \partial x_i} = -b < 0$. Therefore, this payoff function is submodular (and has decreasing differences).

4. In Bertrand oligopoly (Example 1.7, page 8), profit of firm i from setting price x_i when opponents set prices x_{-i} is given by

$$u_i(x_i, x_{-i}) = (x_i - c)\left(a - bx_i + \beta\left(\sum_{j \neq i} x_j\right)\right).$$

For every firm i and for every competitor $j \neq i$, $\frac{\partial^2 u_i}{\partial x_j \partial x_i} = \beta > 0$. Therefore, this payoff function is not submodular (and does not have decreasing differences).

4.1.3 *Dual Single Crossing Property*

For posets $(X, \preceq_X)$ and $(Y, \preceq_Y)$, a function $f : X \times Y \to \mathbb{R}$ has ***dual single crossing property in*** (x, y), if for every $\hat{x} \preceq \tilde{x}$ and for every $\hat{y} \preceq \tilde{y}$, (1) $f(\tilde{x}, \hat{y}) \leq f(\hat{x}, \hat{y}) \Rightarrow f(\tilde{x}, \tilde{y}) \leq f(\hat{x}, \tilde{y})$ and (2) $f(\tilde{x}, \hat{y}) < f(\hat{x}, \hat{y}) \Rightarrow f(\tilde{x}, \tilde{y}) < f(\hat{x}, \tilde{y})$.

Dual single crossing property requires that for each $\hat{x} \preceq \tilde{x}$, if the function $f(\tilde{x}, \cdot) - f(\hat{x}, \cdot)$ crosses zero (from above) at $\hat{y}$, that is, $f(\tilde{x}, \hat{y}) - f(\hat{x}, \hat{y}) \leq 0$, then it can never go above zero again, that is, for every $\tilde{y} \succeq \hat{y}$, $f(\tilde{x}, \tilde{y}) - f(\hat{x}, \tilde{y}) \leq 0$. Moreover, if the function $f(\tilde{x}, \cdot) - f(\hat{x}, \cdot)$ strictly crosses zero (from above) once, it can never go back to zero (or above zero) again. In this sense, this property is dual to single crossing property. The following proposition shows another connection between the two.

Proposition 4.8 *Let* $(X, \preceq_X)$, $(Y, \preceq_Y)$ *be posets and* $f : X \times Y \to \mathbb{R}$. f *has dual single crossing property in* (x, y), *if, and only if,* $-f$ *has single crossing property in* (x, y).

Proof Let $g = -f$ and consider $\hat{x} \preceq_X \tilde{x}$ and $\hat{y} \preceq_Y \tilde{y}$. For sufficiency, suppose f has dual single crossing property in (x, y). Then

$$\begin{aligned} g(\tilde{x}, \hat{y}) \geq g(\hat{x}, \hat{y}) &\Leftrightarrow f(\tilde{x}, \hat{y}) \leq f(\hat{x}, \hat{y}) \\ &\Rightarrow f(\tilde{x}, \tilde{y}) \leq f(\hat{x}, \tilde{y}) \\ &\Leftrightarrow g(\tilde{x}, \tilde{y}) \geq g(\hat{x}, \tilde{y}), \end{aligned}$$

where the implication follows from f having dual single crossing property. The case of strict inequality is proved similarly. For necessity, suppose g has single crossing property in (x, y). Then

$$\begin{aligned} f(\tilde{x}, \hat{y}) \leq f(\hat{x}, \hat{y}) &\Leftrightarrow g(\tilde{x}, \hat{y}) \geq g(\hat{x}, \hat{y}) \\ &\Rightarrow g(\tilde{x}, \tilde{y}) \geq g(\hat{x}, \tilde{y}) \\ &\Leftrightarrow f(\tilde{x}, \tilde{y}) \leq f(\hat{x}, \tilde{y}), \end{aligned}$$

where the implication follows from g having single crossing property. The case of strict inequality is proved similarly. □

This proposition and earlier results for single crossing property show that dual single crossing property is not commutative in its components, that is, f has dual single crossing in (x, y) does not necessarily imply that f has dual single crossing property in (y, x).

Example 4.9 Consider posets $X = Y = \{0, 1\}$ with $0 \prec 1$, and $f : X \times Y \to \mathbb{R}$ given by $f(0, 0) = 3$, $f(1, 0) = 0$, $f(0, 1) = 2$, $f(1, 1) = 1$. Then f has dual single crossing property in (x, y) but not in (y, x).

It is easy to check that if f has decreasing differences in (x, y), then f has dual single crossing property in (x, y). The converse is not necessarily true. The example above shows this as well, because in the example, f has single crossing property in (x, y), but f does not have increasing differences in (x, y).

It follows immediately from the definition that dual single crossing property is invariant to strictly increasing transformations, that is, if f has dual single crossing property in (x, y) and $g : \mathbb{R} \to \mathbb{R}$ is a strictly increasing function, then $g \circ f$ has dual single crossing property in (x, y). In this sense, dual single crossing property is as an ordinal generalization of decreasing differences.

Dual single crossing property for particular components of a product space is formalized in a manner similar to single crossing property and using the same notation. Consider posets $X_1, \ldots, X_N$ and let $X = X_1 \times \cdots \times X_N$. For components i, j with $i \neq j$, a function $f : X \to \mathbb{R}$ has ***dual single crossing property in*** (x_i, x_j), if for every $x_{-(i,j)} \in X_{-(i,j)}$, $f(x_i, x_j, x_{-(i,j)})$ has dual single crossing property in (x_i, x_j). A function $f : X \to \mathbb{R}$ has ***dual single crossing property on*** X, if for every i, j with $i \neq j$, f has dual single crossing property in (x_i, x_j).

4.1.4 *Quasisubmodular Function*

Let $(X, \preceq_X)$ be a lattice. A function $f : X \to \mathbb{R}$ is ***quasisubmodular on*** X, if for every $x, y \in X$, (1) $f(x) \leq f(x \wedge y) \Rightarrow f(x \vee y) \leq f(y)$, and (2) $f(x) < f(x \wedge y) \Rightarrow f(x \vee y) < f(y)$. If x and y are comparable, the condition holds trivially, and therefore, a nontrivial application requires x and y to be noncomparable. Therefore, if X is a chain, then every function $f : X \to \mathbb{R}$ is quasisubmodular. In particular, every function $f : \mathbb{R} \to \mathbb{R}$, is quasisubmodular.

It follows immediately from the definition that f is quasisubmodular on X, if, and only if, $-f$ is quasisupermodular on X.

It is easy to check that if f is submodular on X, then f is quasisubmodular on X. The converse is not necessarily true.

Example 4.10 The function f on pairs of strictly positive real numbers $x > 0$ and $y > 0$ given by $f(x, y) = (x + y)^2$ is quasisubmodular, because it is a strictly increasing transformation of the submodular function $h(x, y) = 2\ln(x + y)$, but f is not submodular, because $\frac{\partial^2 f}{\partial x_2 \partial x_1} = 2 > 0$.

It follows immediately from the definition that quasisubmodular property is invariant to strictly increasing transformations, that is, if f is quasisubmodular on X and $g : \mathbb{R} \to \mathbb{R}$ is a strictly increasing function, then $g \circ f$ is quasisubmodular on X. In this sense, quasisubmodular property is viewed as an ordinal generalization of submodular property.

Proposition 4.11 *Let* $X_1, \ldots, X_N$ *be lattices,* $X = X_1 \times \cdots \times X_N$*, and* $f : X \to \mathbb{R}$.
If f *is quasisubmodular on* X*, then* f *has dual single crossing property on* X.

Proof If f is quasisubmodular on X, then $-f$ is quasisupermodular on X. Using Theorem 3.9 (page 56), it follows that $-f$ has single crossing property on X, and therefore, f has dual single crossing property on X. □

In general, dual single crossing property (even on all pairs of components) does not imply quasisubmodular property, as shown by the following example.

Example 4.12 Let $X_1 = \{a, b, c, d\}$ with $a \prec b \prec d, a \prec c \prec d$, and b and c are not comparable. Let $X_2 = \{0, 1\}$ with $0 \prec 1$ and consider $X = X_1 \times X_2$ with the product order. Let $f : X \to \mathbb{R}$ be given by

$$f(a, 0) = 3,\ f(b, 0) = 2,\ f(c, 0) = 0,\ f(d, 0) = 1,$$
$$f(a, 1) = 3,\ f(b, 1) = 2,\ f(c, 1) = 0,\ f(d, 1) = 1.$$

It is easy to check that f has dual single crossing property in (x_1, x_2) and in (x_2, x_1), but f is not quasisubmodular on X, because if we let $y = (b, 0)$ and $z = (c, 0)$, then $f(y \wedge z) - f(y) = 1 \geq 0$, but $f(z) - f(y \vee z) = -1 \ngeq 0$.

4.1.5 *Decreasing Maximizers*

Let X be a lattice, T a poset, S a subset of X, and $f : X \times T \to \mathbb{R}$. Consider the maximization problem $\max_{x \in S} f(x, t)$ and let

$$B(S, t) = \arg\max_{x \in S} f(x, t)$$

be the solution to the maximization problem. $B(S, t)$ is ***(weakly) increasing in*** S ***and (weakly) decreasing in*** t, if for every $\hat{S} \sqsubseteq \tilde{S}$, $\hat{t} \preceq_T \tilde{t}$, $B(\hat{S}, \tilde{t}) \sqsubseteq B(\tilde{S}, \hat{t})$. The following characterization is due to Roy and Sabarwal (2010). Their proof is expanded to include more detail.

Theorem 4.13 (Decreasing Maximizers Theorem) *Let X be a lattice, T a poset, S a subset of X, $f : X \times T \to \mathbb{R}$, and $B(S, t) = \arg\max_{x \in S} f(x, t)$. $B(S, t)$ is (weakly) increasing in S and (weakly) decreasing in t, if, and only if, f has dual single crossing property in (x, t) and for every $t \in T$, $f(\cdot, t)$ is quasisupermodular on X.*

Proof For sufficiency, suppose $B(S, t)$ is (weakly) increasing in S and (weakly) decreasing in t. To show that f has dual single crossing property in (x, t), consider $\hat{x} \preceq \tilde{x}$ and $\hat{t} \preceq \tilde{t}$. Suppose $f(\hat{x}, \hat{t}) \geq f(\tilde{x}, \hat{t})$. If $\hat{x} = \tilde{x}$ then $f(\hat{x}, \tilde{t}) \geq f(\tilde{x}, \tilde{t})$ is trivially true, so suppose $\hat{x} \prec \tilde{x}$, and let $S = \{\hat{x}, \tilde{x}\}$. In this case, $\hat{x} \in B(S, \hat{t})$, and combined with $B(S, \tilde{t}) \sqsubseteq B(S, \hat{t})$, it follows that $\hat{x} \in B(S, \tilde{t})$, and therefore, $f(\hat{x}, \tilde{t}) \geq f(\tilde{x}, \tilde{t})$, as desired. Now suppose

$f(\hat{x},\hat{t}) > f(\tilde{x},\hat{t})$, which implies that $\hat{x} \prec \tilde{x}$, and let $S = \{\hat{x},\tilde{x}\}$. In this case, $\{\hat{x}\} = B(S,\hat{t})$, and combined with $B(S,\tilde{t}) \sqsubseteq B(S,\hat{t})$, it follows that $B(S,\tilde{t}) = \{\hat{x}\}$, and therefore, $f(\hat{x},\tilde{t}) > f(\tilde{x},\tilde{t})$.

To show that for every $t \in T$, $f(\cdot,t)$ is quasisupermodular on X, fix $t \in T$ and consider $x, y \in X$. Suppose $f(x,t) \geq f(x \wedge y, t)$. Let $\hat{S} = \{x, x \wedge y\}$ and $\tilde{S} = \{y, x \vee y\}$, and notice that $\hat{S} \sqsubseteq \tilde{S}$. In this case, $x \in B(\hat{S},t)$, and combined with $B(\hat{S},t) \sqsubseteq B(\tilde{S},t)$, it follows that $x \vee y \in B(\tilde{S},t)$, and therefore, $f(x \vee y, t) \geq f(y,t)$, as desired. Now suppose $f(x,t) > f(x \wedge y, t)$. In this case, $\{x\} = B(\hat{S},t)$, and combined with $B(\hat{S},t) \sqsubseteq B(\tilde{S},t)$, it follows that $B(\tilde{S},t) = \{x \vee y\}$, and therefore, $f(x \vee y, t) > f(y,t)$.

For necessity, suppose f has dual single crossing property in (x,t) and for every $t \in T$, $f(\cdot,t)$ is quasisupermodular on X, and consider $\hat{S} \sqsubseteq \tilde{S}$ and $\hat{t} \preceq \tilde{t}$. If either $B(\hat{S},\tilde{t})$ is empty or $B(\tilde{S},\hat{t})$ is empty, then the conclusion follows trivially. Otherwise, let $x \in B(\hat{S},\tilde{t})$ and $y \in B(\tilde{S},\hat{t})$. As $\hat{S} \sqsubseteq \tilde{S}$, it follows that $x \wedge y \in \hat{S}$ and $x \vee y \in \tilde{S}$, and therefore,

$$\begin{aligned} x \in B(\hat{S},\tilde{t}) &\Rightarrow f(x,\tilde{t}) \geq f(x \wedge y, \tilde{t}) \\ &\Leftrightarrow f(x,\tilde{t}) \nless f(x \wedge y, \tilde{t}) \\ &\Rightarrow f(x,\hat{t}) \nless f(x \wedge y, \hat{t}) \\ &\Rightarrow f(x \vee y, \hat{t}) \geq f(y,\hat{t}), \end{aligned}$$

where the second implication follows from f having dual single crossing property in (x,t) and the third implication follows from $f(\cdot,\hat{t})$ being quasisupermodular on X. Combined with $y \in B(\tilde{S},\hat{t})$ and $x \vee y \in \tilde{S}$, it follows that $x \vee y \in B(\tilde{S},\hat{t})$. Similarly,

$$\begin{aligned} y \in B(\tilde{S},\hat{t}) &\Rightarrow f(y,\hat{t}) \geq f(x \vee y, \hat{t}) \\ &\Leftrightarrow f(x \vee y,\hat{t}) \ngtr f(y, \hat{t}) \\ &\Rightarrow f(x,\hat{t}) \ngtr f(x \wedge y, \hat{t}) \\ &\Rightarrow f(x, \tilde{t}) \leq f(x \wedge y,\tilde{t}), \end{aligned}$$

where the second implication follows from $f(\cdot,\hat{t})$ being quasisupermodular on X and the third implication follows from f having dual single crossing property in (x,t). Combined with $x \in B(\hat{S},\tilde{t})$ and $x \wedge y \in \hat{S}$, it follows that $x \wedge y \in B(\hat{S},\tilde{t})$. We conclude that $B(\hat{S},\tilde{t}) \sqsubseteq B(\tilde{S},\hat{t})$. □

Corollary 4.14 *If f has decreasing differences in (x, t) and for every $t \in T$, $f(\cdot, t)$ is supermodular on X, then $B(S, t)$ is (weakly) increasing in S and (weakly) decreasing in t.*

Proof As decreasing differences in (x, t) imply dual single crossing property in (x, t) and a supermodular function is quasisupermodular, the result follows from the previous theorem. □

Corollary 4.15 *Let X be a chain, T a poset, S a subset of X, $f : X \times T \to \mathbb{R}$, and $B(S, t) = \arg\max_{x \in S} f(x, t)$.*
$B(S, t)$ is (weakly) increasing in S and (weakly) decreasing in t, if, and only if, f has dual single crossing property in (x, t).

Proof When X is a chain, every function on X is quasisupermodular, and therefore, the condition *for every $t \in T$, $f(\cdot, t)$ is quasisupermodular on X* in the previous theorem is satisfied automatically. □

4.2 Game with Strategic Substitutes

4.2.1 Definition

A game with strategic substitutes models a decentralized and interdependent decision-making environment in which each player i has contradirectional incentives, that is, each player i has an incentive to take a lower action when opponents of i take a higher action.

One way to incorporate contradirectional incentives is to posit that marginal benefit to player i from a higher action decreases when opponents take higher actions. This is formalized by u_i having decreasing differences in (x_i, x_{-i}).

Dual single crossing property is weaker than decreasing differences, and decreasing maximizers theorem shows that dual single crossing property (combined with quasisupermodular property) is sufficient for decreasing best responses. In other words, a sufficient condition for player i to have contradirectional incentives is for player i payoff to have dual single crossing property in (x_i, x_{-i}) and quasisupermodular property in own action.

Strategic substitutes is formalized as follows. In a lattice game $\Gamma = (X_i, u_i)_{i=1}^I$, ***player i has strategic substitutes***, if u_i has dual single crossing property in (x_i, x_{-i}) and for each x_{-i}, $u_i(\cdot, x_{-i})$ is quasisupermodular on X_i.

A ***game with strategic substitutes***, or ***GSS***, is a lattice game in which every player i has strategic substitutes. In other words, a GSS is a lattice game $\Gamma = (X_i, u_i)_{i=1}^I$, where

1. There are finitely many players, indexed by $i \in \{1, \ldots, I\}$.
2. For each i, the action space of player i is X_i, where X_i is a nonempty, subcomplete lattice in $\mathbb{R}^{n_i}$ with the Euclidean order.
3. For each i, payoff for player i is $u_i : X_i \times X_{-i} \to \mathbb{R}$, where

   (a) u_i is upper semicontinuous on X and continuous on X_{-i}, and
   (b) u_i has dual single crossing property in (x_i, x_{-i}) and for each x_{-i}, $u_i(\cdot, x_{-i})$ is quasisupermodular on X_i.

Games with strategic substitutes have been analyzed by many authors. Economic applications using these ideas in two player games are available in Fudenberg and Tirole (1984) and Bulow et al. (1985). Cournot oligopoly is considered in Amir (1996). More general results are available in Zimper (2007), Roy and Sabarwal (2008), Roy and Sabarwal (2010), Roy and Sabarwal (2012), and Monaco and Sabarwal (2016). These include results for more general lattice games as well. Embeddings are considered in Cao et al. (2018). Models with aggregative games include Rosenthal (1973), Monderer and Shapley (1996), Dubey et al. (2006), Jensen (2010), and Acemoglu and Jensen (2013).

Consistent with the treatment for GSC, although a player's action space in a GSS is assumed to be finite dimensional, the material on contradirectional incentives in the previous section is developed in its natural and more general setting. Moreover, proofs of results in this section are developed in a manner that shows their natural extension to more general settings with additional assumptions.

The dove hawk game (Example 1.2, page 3), team trust game (Example 1.3, page 4), multi-player externality game (Example 1.6, page 7), and Cournot oligopoly (Example 1.8, page 9) are all games with strategic substitutes. As can be seen from the information in Examples 4.2 (page 84) and Examples 4.7 (page 87), in each of these cases, the payoff of each player has decreasing differences, which implies that payoff of each player satisfies dual single crossing property. Moreover, in each of these cases, the action space of each player is a chain, and therefore, each player's payoff is quasisupermodular on their own action space.

4.2.2 Decreasing Best Response

Recall that in a lattice game best response set of player i to opponent actions x_{-i} is $B_i(x_{-i}) = \arg\max_{x_i \in X_i} u_i(x_i, x_{-i})$, the best response correspondence for player i is $B_i : X_{-i} \rightrightarrows X_i$, mapping x_{-i} to $B_i(x_{-i})$, and the joint best response correspondence is $B : X \rightrightarrows X$, given by $B(x) = B_1(x_{-1}) \times B_2(x_{-2}) \times \cdots \times B_I(x_{-I})$.

At the more general level of a lattice game, the individual and joint best response correspondence are nonempty valued, compact valued, and upper hemicontinuous (Theorem 2.26, page 32). The decreasing maximizers theorem combined with characterization of subcomplete lattices in Theorem 2.14 (page 22) show that they are subcomplete valued and decreasing as well.

Theorem 4.16 *In every lattice game* $\Gamma = (X_i, u_i)_{i=1}^I$,

1. *If player i has strategic substitutes, then best response correspondence for player i is nonempty valued, compact valued, subcomplete valued, upper hemicontinuous, and (weakly) decreasing.*
2. *If the game is a GSS, then joint best response correspondence is nonempty valued, compact valued, subcomplete valued, upper hemicontinuous, and (weakly) decreasing.*

Proof For statement (1), let B_i denote the best response correspondence for player i. Theorem 2.26 (page 32) shows that B_i is nonempty valued, compact valued, and upper hemicontinuous. When player i has strategic substitutes, Corollary 3.16 (page 61) shows that for every x_{-i}, $B_i(x_{-i})$ is a sublattice, and therefore, by Theorem 2.14 (page 22), $B_i(x_{-i})$ is subcomplete in X_i. Moreover, Theorem 4.13 (page 91) implies that B_i is (weakly) decreasing. This proves statement (1).

For statement (2), let B denote the joint best response correspondence. Theorem 2.26 (page 32) shows that B is nonempty valued, compact valued, and upper hemicontinuous. Statement (1) implies that for every i, $B_i(x_{-i})$ is subcomplete in X_i, and therefore, for every $x \in X$, $B(x)$ is subcomplete. Moreover, for every i, B_i is (weakly) decreasing in x_{-i}, and therefore, B is (weakly) decreasing in x. □

As shown in Examples 2.27 (page 33), in dove hawk game, team trust game, multi-player externality game, and Cournot oligopoly, the best response correspondence for each player is (weakly) decreasing, and therefore, the joint best response correspondence in the game is (weakly) decreasing.

A (weakly) decreasing best response correspondence implies behavioral patterns in a GSS that are different from a GSC. In particular, using the same version of best response dynamics as in GSC is not helpful here, because if you start from inf X then best response implies that everyone takes very high actions and in turn, best response to very high actions is to take very low actions. Therefore, the same definition as in GSC will lead to high and low cycles. A more useful construction is to take mixtures of directionally extremal best response dynamics from inf X and sup X. This is more complex than in GSC, but turns out to have important information to study ranges of outcomes in GSS.

In a GSS, the ***best response dynamic from*** inf X is the sequence $(y^n)_{n=0}^{\infty}$ in X, where $y^0 = \inf X$, and for $n \geq 0$,

$$y^n = \begin{cases} \overline{B}(y^{n-1}) & \text{if } n \text{ is odd, and} \\ \underline{B}(y^{n-1}) & \text{if } n \text{ is even.} \end{cases}$$

Here, $\overline{B}(y^{n-1}) = \sup B(y^{n-1})$ and $\underline{B}(y^{n-1}) = \inf B(y^{n-1})$. If best responses are singleton valued, this definition is the same as the one given for GSC. More generally, we may view this as directionally extremal best response dynamics in the following sense. For $y^0 = \inf X$, as y^0 is the smallest profile of actions, strategic substitutes implies that $B(y^0)$ are "high" best responses, and in this case, we take the highest of these best responses, that is, $y^1 = \overline{B}(y^0)$. As y^1 is a high profile of actions, $B(y^1)$ are "low" best responses, and in this case, we take the lowest of these best responses, that is, $y^2 = \underline{B}(y^1)$, and so on. Similarly, the ***best response dynamic from*** sup X is the sequence $(z^n)_{n=0}^{\infty}$, where $z^0 = \sup X$, and for $n \geq 0$,

$$z^n = \begin{cases} \underline{B}(z^{n-1}) & \text{if } n \text{ is odd, and} \\ \overline{B}(z^{n-1}) & \text{if } n \text{ is even.} \end{cases}$$

The ***lower mixture of*** (y^n) ***and*** (z^n) as the sequence $(\underline{x}^n)_{n=0}^{\infty}$, where $\underline{x}^0 = \inf X$, and for $n \geq 1$,

$$\underline{x}^n = \begin{cases} z^n & \text{if } n \text{ is odd, and} \\ y^n & \text{if } n \text{ is even.} \end{cases}$$

The ***upper mixture of*** (y^n) ***and*** (z^n) is the sequence $(\overline{x}^n)_{n=0}^{\infty}$, where $\overline{x}^0 = \sup X$ and for $n \geq 1$,

$$\overline{x}^n = \begin{cases} y^n & \text{if } n \text{ is odd, and} \\ z^n & \text{if } n \text{ is even.} \end{cases}$$

The lower and upper mixtures turn out to be analytically useful. This is seen in the following result. Notably, upper hemicontinuity of best response correspondence is available from the more general result for lattice games.

Theorem 4.17 *In a GSS, let* $(y^n), (z^n)$ *be the best response dynamics from* $\inf X$ *and* $\sup X$*, respectively, and let* $(\underline{x}^n)_{n=0}^{\infty}$ *and* $(\overline{x}^n)_{n=0}^{\infty}$ *be their lower and upper mixtures, respectively.*

1. *For every* n, $\underline{x}^n \preceq \underline{x}^{n+1} \preceq \overline{x}^{n+1} \preceq \overline{x}^n$.
2. *Let* $\underline{x} = \bigvee_{n=0}^{\infty} \underline{x}^n$ *and* $\overline{x} = \bigwedge_{n=0}^{\infty} \overline{x}^n$. *Then* $\underline{x} \preceq \overline{x}$.
3. $\underline{x} \in B(\overline{x})$ *and* $\overline{x} \in B(\underline{x})$.

Proof For statement (1), we use induction. The result is true for $n = 0$, as $\underline{x}^0 = \inf X \preceq \sup X = \overline{x}^0$, and therefore, strategic substitutes implies $B(\overline{x}^0) \sqsubseteq B(\underline{x}^0)$, and consequently, $\underline{x}^0 \preceq \underline{x}^1 = \underline{B}(z^0) = \underline{B}(\overline{x}^0) \preceq \overline{B}(\underline{x}^0) = \overline{B}(y^0) = \overline{x}^1 \preceq \overline{x}^0$. Suppose $\underline{x}^n \preceq \underline{x}^{n+1} \preceq \overline{x}^{n+1} \preceq \overline{x}^n$. As case 1, suppose n is odd. Then we have that $\underline{x}^n = z^n$, $\underline{x}^{n+1} = y^{n+1}$, $\overline{x}^n = y^n$, and $\overline{x}^{n+1} = z^{n+1}$. Therefore, by strategic substitutes, $B(y^n) \sqsubseteq B(z^{n+1}) \sqsubseteq B(y^{n+1}) \sqsubseteq B(z^n)$. In particular, $y^{n+1} = \underline{B}(y^n) \preceq z^{n+2} = \underline{B}(z^{n+1}) \preceq y^{n+2} = \overline{B}(y^{n+1}) \preceq z^{n+1} = \overline{B}(z^n)$. The result follows, because $\underline{x}^{n+1} = y^{n+1}$, $\underline{x}^{n+2} = z^{n+2}$, $\overline{x}^{n+2} = y^{n+2}$, and $\overline{x}^{n+1} = z^{n+1}$. The case where n is even follows by a similar analysis. This proves statement (1).

For statement (2), statement (1) shows that $(\underline{x}^n)_{n=1}^{\infty}$ is a (weakly) increasing sequence and $(\overline{x}^n)_{n=1}^{\infty}$ is a (weakly) decreasing sequence and both sequences are in a subcomplete lattice. Therefore, $\underline{x}$ and $\overline{x}$ exist and are well defined. As $\underline{x}^n \preceq \overline{x}^n$ for every n, it follows that $\underline{x} \preceq \overline{x}$.

For statement (3), let $\xi^n = \underline{x}^{n+1}$. Then $(\overline{x}^n)_{n=0}^{\infty}$ and $(\xi^n)_{n=0}^{\infty}$ are sequences in X, $\overline{x}^n \to \overline{x}$, $\xi^n \to \underline{x}$, and for every n, $\xi^n \in B(\overline{x}^n)$. As B is upper hemicontinuous, it follows that $\underline{x} \in B(\overline{x})$. Similarly, it can be shown that $\overline{x} \in B(\underline{x})$. □

Statement (3) in this theorem shows that limits of upper and lower mixtures of best response dynamics are best responses to each other. This does not necessarily imply that they are Nash equilibria, as shown by the three firm Cournot oligopoly (Example 4.33, page 112). Nevertheless, these profiles have behavioral content in terms of rationalizable strategies as follows.

In a GSS, a profile of actions $x \in X$ is ***simply rationalizable***, if there is $y \in X$ such that for every player i, $x_i \in B_i(y_{-i})$ and $y_i \in B_i(x_{-i})$. In this case, we say that x is simply rationalizable using y. Symmetry in the definition means that x is simply rationalizable using y, if, and only if, y is simply rationalizable using x. Intuitively, a profile of actions x is simply rationalizable if there is a profile of actions y such that every player i can rationalize playing x_i with a short cycle of conjectures using y as follows. Player i plays x_i because they believe their opponents will play y_{-i}, because each opponent j believes further that others are playing x_{-j}.

It is immediate from the definition that a Nash equilibrium x is simply rationalizable, because we may choose $y = x$. Simply rationalizable actions allow for additional outcomes as long as they are justifiable with short cycles of reasoning. This may be viewed as a type of bounded rationality. Theorem 2.29 (page 39) shows that simply rationalizable profiles are serially undominated.

Applying the definition to limits of lower and upper mixtures of best response dynamics, it follows immediately that both $\underline{x}$ and $\overline{x}$ are simply rationalizable using each other.

Theorem 4.18 *In a GSS, let $\underline{x}$, $\overline{x}$ be the respective limit of upper and lower mixtures of best response dynamics, as above. The following are equivalent.*

1. *The best response dynamic from* inf X *converges.*
2. *The best response dynamic from* sup X *converges.*
3. $\underline{x} = \overline{x}$

Proof To demonstrate equivalence of (1) and (3), notice that for every $n \geq 0$,

$$y^n = \begin{cases} \overline{x}^n & \text{if } n \text{ is odd, and} \\ \underline{x}^n & \text{if } n \text{ is even.} \end{cases}$$

If (y^n) converges, then its odd and even subsequences must necessarily converge to the same value. As its odd subsequence is a subsequence of the convergent sequence $(\overline{x}^n)$ and its even subsequence is a subsequence of the convergent sequence $(\underline{x}^n)$, it follows that $\underline{x} = \overline{x}$. In the other direction, if $\underline{x} = \overline{x}$, then as the sequence (y^n) is composed of terms that converge to this common value, it follows that (y^n) converges to this common value. The equivalence of (2) and (3) is proved similarly. □

Notice that statement (1) only requires that the best response dynamic from inf X converges. Knowledge of its limit is not required. Therefore, it is sufficient to show that the best response dynamic from inf X is a convergent sequence. This may be easier to show in practice, perhaps by showing it is a Cauchy sequence or by showing that it satisfies a contraction mapping argument.

Moreover, if it is demonstrated that the best response dynamic from inf X converges, then it automatically ensures that the best response dynamic from sup X converges as well. In fact, they must converge to the same value ($\underline{x} = \overline{x}$).

This is in contrast to the case for GSC. In a GSC, the best response dynamic from inf X always converges and the best response dynamic from sup X always converges, but their limits are not guaranteed to be the same. In a GSS, if one of these dynamics converges, then the other one must converge as well, and to the same limit. If one of these dynamics does not converge, then the other one does not converge either.

4.2.3 *Nash Equilibrium Set*

A central difference between GSC and GSS is that the Nash equilibrium set in a GSS may be empty.

Example 4.19 Consider a three player, two action game with payoffs given in Table 4.1 (page 100). With $L \prec H$, it is easy to check that this is a GSS. It is also easy to check that there is no Nash equilibrium in this game.

In GSC, the Nash equilibrium set is always a nonempty, complete lattice (Theorem 3.21, page 66). In GSS, the Nash equilibrium set may be empty, as shown first in Roy and Sabarwal (2012).

This example shows a limitation of efforts to transform a GSS into a GSC by reversing the order on the action space of some players. This is a well-known trick in the case of a two player GSS. As Example 4.19 (page 99) shows, this may fail in the simplest possible setting beyond two player games, that is, in a three player, two action game.

More generally, Cao et al. (2018) study the problem of embedding one game into another game in the sense that the game being embedded is a

Table 4.1 GSS with no Nash equilibrium

P3 = L:

P1 \ P2	L	H
L	2, 0, 0	0, 3, 3
H	3, 3, 1	1, 1, 3

P3 = H:

P1 \ P2	L	H
L	2, 2, 2	3, 5, 1
H	0, 2, 3	1, 0, 1

component of the larger game and restriction of the Nash equilibrium set of the larger game to this component yields the Nash equilibrium set of the game being embedded. They show that it is always possible to embed a GSC into a (larger) GSS, but in general, it is impossible to embed a GSS into a GSC in the same manner.

With additional assumptions, the Nash equilibrium set in a GSS is nonempty without resorting to change in order. As mentioned in Sect. 2.2.2 (page 35), there are several classes of lattice games in which the Nash equilibrium set is nonempty. As games with strategic substitutes are lattice games, when a GSS falls in one or more of those classes, it has a Nash equilibrium.

Examples 4.20 Consider the following examples.

1. In the dove hawk game (Example 1.2, page 3), the Nash equilibrium set is $\mathcal{E} = \{(D, H), (H, D)\}$. Each Nash equilibrium is strict and neither is comparable to the other in the underlying partial order.
2. The team trust game (Example 1.4, page 5) has a unique Nash equilibrium, $\mathcal{E} = \{(D, D)\}$.
3. The multi-player externality game (Example 1.6, page 7) has I Nash equilibria, given by the basis vectors in $\mathbb{R}^I$.

$$\mathcal{E} = \{(1, 0, \ldots, 0), (0, 1, \ldots, 0), \ldots, (0, 0, \ldots, 1)\} \subset \{0, 1\}^I.$$

   Each Nash equilibrium is strict and the Nash equilibrium set is totally unordered, that is, no pair of Nash equilibria are comparable in the underlying partial order.
4. In Cournot oligopoly (Example 1.8, page 9), the Nash equilibrium is unique, strict, and symmetric, with equilibrium output of each firm i given by $x_i^* = \frac{a-c}{I+1}$. The Nash equilibrium set is trivially a complete lattice.

In these examples, when there are multiple Nash equilibria, no Nash equilibrium is comparable to the others in the underlying partial order. That is, the Nash equilibrium set is totally unordered.

These examples show another central difference between GSS and GSC. In a GSC the Nash equilibrium set is a nonempty, complete lattice, and therefore, with multiple equilibria, for every pair of Nash equilibria there is one that is (weakly) larger than either of these and another that is (weakly) smaller than either of these. Moreover, there is always a (weakly) smallest Nash equilibrium and a (weakly) largest Nash equilibrium.

In GSS, under mild conditions on one or two players only, the Nash equilibrium set is totally unordered. In other words, the order structure of the Nash equilibrium set is destroyed completely in the sense that no two Nash equilibria are comparable in the underlying partial order. Therefore, with multiple equilibria, it is impossible to have a smallest equilibrium or a largest equilibrium, because those would necessarily be ordered. This is true more generally for lattice games that are not GSC in the following sense.

Let X be a poset and A, B subsets of X. A is ***completely lower than*** B, denoted $A \sqsubset_c B$, if for every $a \in A$ and for every $b \in B$, $a \prec b$. In a lattice game, player i has ***strict strategic substitutes***, if for every $\hat{x}_{-i}$ and $\tilde{x}_{-i}$ in X_{-i}, $\hat{x}_{-i} \prec \tilde{x}_{-i}$ implies $B_i(\tilde{x}_{-i}) \sqsubset_c B_i(\hat{x}_{-i})$. In other words, player i has strict strategic substitutes, if B_i is decreasing in the completely lower than set order. If $B_i(\cdot)$ is singleton valued, then the statement that player i has strict strategic substitutes is equivalent to the statement that $B_i(\cdot)$ is a strictly decreasing function of x_{-i}.

Theorem 4.21 *In every lattice game,*

1. *If one player has strict strategic substitutes and singleton valued best response, or*
2. *If two players have strict strategic substitutes,*

Then the Nash equilibrium set is totally unordered.

Proof If the Nash equilibrium set is empty or a singleton, the conclusion is trivially true.

Suppose condition (1) is satisfied, without loss of generality, for player 1. Consider two distinct Nash equilibria $\hat{x}$ and $\tilde{x}$, and suppose $\hat{x}$ and $\tilde{x}$ are comparable, with $\hat{x} \prec \tilde{x}$. Then either $\hat{x}_{-1} \prec \tilde{x}_{-1}$, or $\hat{x}_{-1} = \tilde{x}_{-1}$ and $\hat{x}_1 \prec \tilde{x}_1$.

If $\hat{x}_{-1} \prec \tilde{x}_{-1}$, then condition (1) implies that $\{\tilde{x}_1\} = B_1(\tilde{x}_{-1}) \sqsubset_c B_1(\hat{x}_{-1}) = \{\hat{x}_1\}$, and therefore, $\tilde{x}_1 \prec \hat{x}_1$, and this contradicts $\hat{x} \prec \tilde{x}$. If $\hat{x}_{-1} = \tilde{x}_{-1}$ and $\hat{x}_1 \prec \tilde{x}_1$, then $\{\tilde{x}_1\} = B_1(\tilde{x}_{-1}) = B_1(\hat{x}_{-1}) = \{\hat{x}_1\}$, and therefore, $\tilde{x}_1 = \hat{x}_1$, and this contradicts $\hat{x}_1 \prec \tilde{x}_1$.

If condition (2) is satisfied, then the joint best response correspondence is never increasing. This can be seen as follows. Suppose condition (2) is satisfied, without loss of generality, for players 1 and 2. Consider $x \prec y$. Then either $\hat{x}_{-1} \prec \tilde{x}_{-1}$, or $\hat{x}_{-1} = \tilde{x}_{-1}$ and $\hat{x}_1 \prec \tilde{x}_1$. If $\hat{x}_{-1} \prec \tilde{x}_{-1}$, then $B_1(y_{-1}) \sqsubset_c B_1(x_{-1})$, and therefore, for every $x_1' \in B_1(x_{-1})$ and for every $y_1' \in B_1(y_{-1})$, $y_1' \prec x_1'$. This implies that for every $x' \in B(x)$ and for every $y' \in B(y)$, $x' \npreceq y'$, and therefore, $x' \nprec y'$. On the other hand, if $x_{-1} = y_{-1}$ and $x_1 \prec y_1$, then $x_{-2} \prec y_{-2}$, and applying the same analysis to player 2, it follows that for every $x' \in B(x)$ and for every $y' \in B(y)$, $x' \npreceq y'$, and therefore, $x' \nprec y'$. It follows that the joint best response correspondence is never increasing. Theorem 2.21 (page 28) now implies that the Nash equilibrium set is totally unordered. □

Statement (1) in this theorem is due to Monaco and Sabarwal (2016). It is useful when best response is unique (singleton valued).

Statement (2) is new. It is useful when best response is a correspondence. Statement (2) shows that in an arbitrary lattice game in which two players have strict strategic substitutes, the joint best response correspondence is never increasing. No conditions are needed on payoffs for other players. Other players may have strategic complements, or strategic substitutes, or neither, and in each case, the joint best response correspondence is never increasing.

The scope of Theorem 4.21 (page 101) is lattice games (not the smaller class of GSS). As a special case, under its assumptions, the Nash equilibrium set in every GSS is totally unordered. The theorem also implies that if we start with a GSC and add one player with a strictly decreasing best response function, the complete lattice structure of the equilibrium set is destroyed completely. More generally, the theorem imposes requirements on one or two players only, and no assumptions about payoffs for other players. Other players may have strategic complements, or strategic substitutes, or neither, and in each case, the equilibrium set is totally unordered.

Example 4.22 (Cournot duopoly with spillover) Consider a Cournot duopoly with inverse market demand given by $p = 20 - \frac{1}{2}(x_1 + x_2)$. Marginal cost for firm 1 is constant at 1 per unit. Profit of firm 1 is given by $\pi_1(x_1, x_2) = (20 - \frac{1}{2}(x_1 + x_2))x_1 - x_1$. Marginal cost for firm 2 is also 1 per unit, but it is affected by an additional externality depending on output of firm 1. For example, firm 1 operation may allow for cheaper access to inputs of production or generate open access technological benefit. This spillover is given by a function $s(x_1)$ that depends on output of firm 1. Profit of firm 2 is given by $\pi_2(x_1, x_2) = (20 - \frac{1}{2}(x_1 + x_2))x_2 - s(x_1)x_2$. Suppose spillover is quadratic and given by $s(x_1) = x_1^2 - \frac{21}{2}x_1 + 11$.

Best response of firm 1 is given by $B_1(x_2) = 20 - \frac{1}{2}x_2$ and best response of firm 2 is given by $B_2(x_1) = \max\{10 + 10x_1 - x_1^2, 0\}$. These are shown in Fig. 4.1 (page 103). Firm 1 has strict strategic substitutes and singleton valued best response. Firm 2 has neither strategic complements nor strategic substitutes. There are three Nash equilibria, $\mathcal{E} = \{(3.55, 32.90), (8.45, 23.10), (20, 0)\}$. The Nash equilibrium set is totally unordered.

When all players have strategic substitutes, (and perhaps, no one has strict strategic substitutes,) the following new result is proved for the set of strict Nash equilibria.

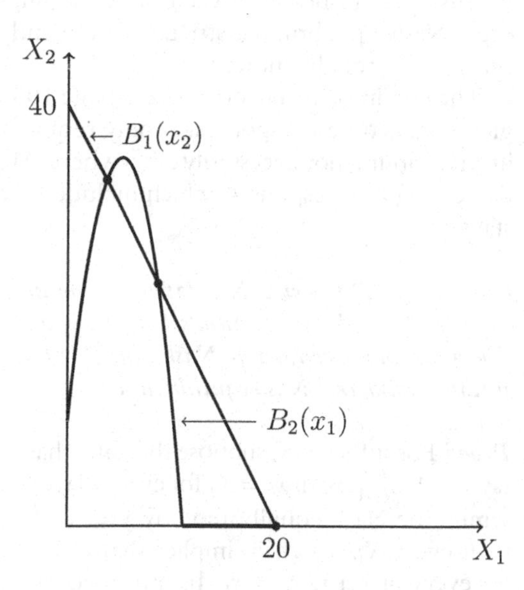

Fig. 4.1 Cournot duopoly with spillovers

Theorem 4.23 *In every GSS, the strict Nash equilibrium set is totally unordered.*

Proof Consider two distinct strict Nash equilibria x^* and y^*. Suppose x^* and y^* are comparable and $x^* \prec y^*$. Let i be such that $x_i^* \prec y_i^*$. Then $x_{-i}^* \preceq y_{-i}^*$. By strategic substitutes and the fact that x^* and y^* are strict Nash equilibria, $\{y_i^*\} = B_i(y_{-i}^*) \sqsubseteq B_i(x_{-i}^*) = \{x_i^*\}$. Thus, $y_i^* \preceq x_i^*$, a contradiction. □

This theorem holds when each player has strategic substitutes. It does not require that even one player has strict strategic substitutes. For example, in the multi-player externality game (Example 1.6, page 7), the best response of each player i is given by

$$B_i(x_{-i}) = \begin{cases} \{1\} & \text{if } \sum_{j \neq i} x_j = 0, \\ \{0\} & \text{if } \sum_{j \neq i} x_j > 0. \end{cases}$$

As $\sum_{j \neq i} x_j > 0$, if, and only if $x_{-i} \neq 0$, it follows that for every player i, B_i is constant for every $x_{-i} \neq 0$. Therefore, no player satisfies the conditions in Theorem 4.21 (page 101). Nevertheless, Theorem 4.23 (page 103) applies, because best response of each player is singleton valued, and therefore, every Nash equilibrium is strict. Consequently, the Nash equilibrium set in this game is totally unordered.

The condition in Theorem 4.23 (page 103) is weaker than requiring singleton valued best responses. It only requires that best response is unique in equilibrium, not necessarily everywhere. Moreover, it allows for (weakly) decreasing best responses, which include best responses that may be constant.

Corollary 4.24 *Let Γ be a lattice game in which every X_i is a chain, and suppose the Nash equilibrium set is totally unordered.*
The game has a symmetric Nash equilibrium, if, and only if, the game has a unique symmetric Nash equilibrium.

Proof For sufficiency, suppose the game has a symmetric Nash equilibrium, say $x = (x)_{i=1}^I$, with $x_i = x_j$ for every player i and j. If the game has another symmetric Nash equilibrium, say $y = (y)_{i=1}^I$, with $y_i = y_j$ for every i, j, then every X_i is a chain implies that either for every player i, $x_i \preceq y_i$, or for every player i, $y_i \preceq x_i$. In other words, either $x \preceq y$ or $y \preceq x$, and in either case, this contradicts that Nash equilibrium set is totally unordered. Necessity is true trivially. □

This result is formulated for lattice games. This presents another contrast to GSC, where it is easy to have multiple symmetric Nash equilibria that are comparable. The coordination game (Example 1.1, page 2), multi-player coordination game (Example 1.5, page 7), and discrete Bertrand duopoly (Example 3.28, page 75) all have this feature. The following result is for GSS.

Corollary 4.25 *Let Γ be a GSS in which every X_i is a chain.*
The game has a strict Nash equilibrium that is symmetric, if, and only if, the game has a unique strict Nash equilibrium that is symmetric.

Proof By Theorem 4.23 (page 103), the strict Nash equilibrium set is totally unordered, and the result follows from the same argument as in Corollary 4.24 (page 104). □

4.2.4 Dominance Solvability and Global Stability

As in GSC, it is useful to first develop bounds for the set of undominated responses at the level of an individual player. Joint behavior can then be derived from behavior of different types of individual players in the game.

Theorem 4.26 *In every GSS,*

1. *For every i and for every $x_{-i}, y_{-i} \in X_{-i}$ with $x_{-i} \preceq y_{-i}$,*

$$[\mathcal{U}_i([x_{-i}, y_{-i}])] = [\underline{B}_i(y_{-i}), \overline{B}_i(x_{-i})].$$

2. *For every $x, y \in X$ with $x \preceq y$, $[\mathcal{U}([x, y])] = [\underline{B}(y), \overline{B}(x)]$.*

Proof For statement (1), consider player i and $x_{-i}, y_{-i} \in X_{-i}$ with $x_{-i} \preceq y_{-i}$. In one direction, as both $\underline{B}_i(y_{-i})$ and $\overline{B}_i(x_{-i})$ are best responses, they are both undominated responses to $[x_{-i}, y_{-i}]$, and therefore, both $\underline{B}_i(y_{-i})$ and $\overline{B}_i(x_{-i})$ are in $\mathcal{U}_i([x_{-i}, y_{-i}])$, whence $[\underline{B}_i(y_{-i}), \overline{B}_i(x_{-i})] \subset [\mathcal{U}_i([x_{-i}, y_{-i}])]$.

In the other direction, suppose $\hat{x}_i \in \mathcal{U}_i([x_{-i}, y_{-i}])$ and $\hat{x}_i \notin [\underline{B}_i(y_{-i}), \overline{B}_i(x_{-i})]$. As case 1, suppose $\hat{x}_i \npreceq \overline{B}_i(x_{-i})$. In this case, $\overline{B}_i(x_{-i}) \vee \hat{x}_i \notin$

$B_i(x_{-i})$. Therefore, for arbitrary $\tilde{x}_{-i} \in [x_{-i}, y_{-i}]$,

$$\begin{aligned} & u_i(\overline{B}_i(x_{-i}) \vee \hat{x}_i, x_{-i}) < u_i(\overline{B}_i(x_{-i}), x_{-i}) \\ \Rightarrow \quad & u_i(\hat{x}_i, x_{-i}) < u_i(\overline{B}_i(x_{-i}) \wedge \hat{x}_i, x_{-i}) \\ \Rightarrow \quad & u_i(\hat{x}_i, \tilde{x}_{-i}) < u_i(\overline{B}_i(x_{-i}) \wedge \hat{x}_i, \tilde{x}_{-i}), \end{aligned}$$

where the first implication follows from u_i being quasisupermodular on X_i, and the second implication follows from u_i having dual single crossing property in (x_i, x_{-i}). As $\tilde{x}_{-i}$ is arbitrary in $[x_{-i}, y_{-i}]$, it follows that $\overline{B}_i(x_{-i}) \wedge \hat{x}_i$ strictly dominates $\hat{x}_i$, which contradicts $\hat{x}_i \in \mathcal{U}_i([x_{-i}, y_{-i}])$. Similarly, case 2, in which $\underline{B}_i(y_{-i}) \not\preceq \hat{x}_i$ yields a contradiction. Therefore, $[\mathcal{U}_i([x_{-i}, y_{-i}])] \subset [\underline{B}_i(y_{-i}), \overline{B}_i(x_{-i})]$.

For statement (2), consider $x, y \in X$ with $x \preceq y$. Statement (1) shows that for every i, $[\mathcal{U}_i([x_{-i}, y_{-i}])] = [\underline{B}_i(y_{-i}), \overline{B}_i(x_{-i})]$, and therefore, $\inf \mathcal{U}_i([x_{-i}, y_{-i}]) = \underline{B}_i(y_{-i}) \in \mathcal{U}_i([x_{-i}, y_{-i}])$ and $\sup \mathcal{U}_i([x_{-i}, y_{-i}]) = \overline{B}_i(x_{-i}) \in \mathcal{U}_i([x_{-i}, y_{-i}])$. Therefore,

$$\begin{aligned} [\mathcal{U}([x, y])] &= \times_{i=1}^{I} [\mathcal{U}_i([x_{-i}, y_{-i}])] \\ &= \times_{i=1}^{I} [\underline{B}_i(y_{-i}), \overline{B}_i(x_{-i})] \\ &= [\underline{B}(y), \overline{B}(x)], \end{aligned}$$

where the first and third equalities follow from Proposition 2.9 (page 20), and the second equality from statement (1). □

Roy and Sabarwal (2012) prove statement (2) in the previous theorem directly. The proof here isolates and proves the individual player argument separately as statement (1). Statement (2) combines the result from statement (1) for all players in the game by using products of generated intervals given in Theorem 2.9 (page 20). Separation of the two in this theorem and in Theorem 3.22 (page 68) makes the proof of the more general case with both types of players (Theorem 5.9, page 142) immediate and intuitive.

Combined with earlier results on behavior of best response dynamics, the following results due to Roy and Sabarwal (2012) for GSS follow in a straightforward manner. Recall that a sequence $(x^n)_{n=0}^{\infty}$ in X is an adaptive dynamic, if for every $n \geq 0$, there is $k_n > n$, such that for every $k \geq k_n$, $x^k \in [\mathcal{U}[P(n,k)]]$, where $P(n,k) = \{x^m \mid n \leq m < k\}$ is play from n to k (not including k). A lattice game is globally stable if there is $x^* \in X$ such that every adaptive dynamic converges to x^*.

Theorem 4.27 *In a GSS, let* $(\underline{x}^n)_{n=1}^{\infty}$ *be the lower mixture of best response dynamics from* $\inf X$ *and* $\sup X$, $(\overline{x}^n)_{n=1}^{\infty}$ *be the upper mixture, and* $\underline{x}$ *and* $\overline{x}$ *be their respective limit.*

1. *Every profile of serially undominated strategies is in the interval* $[\underline{x}, \overline{x}]$.
2. $\underline{x}$ *is the smallest profile of serially undominated strategies in the game, and* $\overline{x}$ *is the largest profile of serially undominated strategies in the game.*
3. $\underline{x}$ *is the smallest simply rationalizable profile and* $\overline{x}$ *is the largest simply rationalizable profile.*
4. *Every adaptive dynamic* $(x^n)_{n=0}^{\infty}$ *satisfies* $\underline{x} \preceq \liminf x^k \preceq \limsup x^k \preceq \overline{x}$.

Proof For statement (1), Theorem 4.26 (page 105) and definition of upper mixture and lower mixture imply that for every $n \geq 0$,

$$[\mathcal{U}([\underline{x}^n, \overline{x}^n])] = [\underline{B}(\overline{x}^n), \overline{B}(\underline{x}^n)] = [\underline{x}^{n+1}, \overline{x}^{n+1}],$$

and therefore, the set of serially undominated strategies satisfies

$$\mathcal{U}^{\infty} \subset \bigcap_{n=0}^{\infty} [\mathcal{U}([\underline{x}^n, \overline{x}^n])] = \bigcap_{n=0}^{\infty} [\underline{x}^{n+1}, \overline{x}^{n+1}] = [\underline{x}, \overline{x}].$$

Statement (2) follows, because $\underline{x} \in B(\overline{x})$ and $\overline{x} \in B(\underline{x})$ imply that $\underline{x}$ and $\overline{x}$ are profiles of serially undominated strategies, and combined with statement (1), $\underline{x}$ is the smallest profile of serially undominated strategies, and $\overline{x}$ is the largest.

Statement (3) follows, because $\underline{x} \in B(\overline{x})$ and $\overline{x} \in B(\underline{x})$ imply that both $\underline{x}$ and $\overline{x}$ are simply rationalizable and Theorem 2.29 (page 39) shows that every simply rationalizable profile is serially undominated.

Statement (4) is proved by first proving that for every $n \geq 0$, there is $k_n > n$, for every $k \geq k_n$, $x^k \in [\underline{x}_n, \overline{x}_n]$. This is proved using induction on n. For $n = 0$, this trivially true, because $[\underline{x}^0, \overline{x}^0] = X$. Suppose this statement is true for fixed n. Let $k_n > n$ be such that for every $k \geq k_n$, $x^k \in [\underline{x}^n, \overline{x}^n]$. Then for every $k > k_n$, $P(k_n, k) \subset [\underline{x}^n, \overline{x}^n]$. Moreover, applying the definition of adaptive dynamics to index k_n, let $k_{n+1} > k_n$ be such that for every $k \geq k_{n+1}$, $x^k \in [\mathcal{U}([P(k_n, k)])]$. Notice that for every $k \geq k_{n+1}$, $P(k_n, k) \subset [\underline{x}^n, \overline{x}^n]$. Therefore, $k_{n+1} > n + 1$, and for every $k \geq k_{n+1}$,

$$x^k \in [\mathcal{U}([P(k_n, k)])] \subset [\mathcal{U}([\underline{x}^n, \overline{x}^n])] = [\underline{x}^{n+1}, \overline{x}^{n+1}].$$

Statement (4) now follows, because for every $n \geq 0$,

$$\underline{x}^n \preceq \liminf x^k \preceq \limsup x^k \preceq \overline{x}^n,$$

and therefore, $\underline{x} \preceq \liminf x^k \preceq \limsup x^k \preceq \overline{x}$. □

Theorem 4.28 *In every GSS, the following are equivalent.*

1. *The game is globally stable.*
2. *The game is dominance solvable.*
3. *The game has a unique simply rationalizable profile.*
4. $\underline{x} = \overline{x}$
5. *The best response dynamic from* $\inf X$ *converges.*
6. *The best response dynamic from* $\sup X$ *converges.*

Proof The equivalence of statements (2) and (4) follows from statements (1) and (2) of Theorem 4.27 (page 106). The equivalence of statements (3) and (4) follows from statement (3) of Theorem 4.27 (page 106). Statement (4) implies statement (1), using statement (4) of Theorem 4.27 (page 106). Statement (1) implies statement (4), because if every adaptive dynamic converges to the same point, then the best response dynamic from $\inf X$ and the best response dynamic from $\sup X$, both of which are adaptive dynamics, converge to the same point, and therefore, $\underline{x} = \overline{x}$. The equivalence of statements (4), (5), and (6) is proved in Theorem 4.18 (page 98). □

This result shows that in GSS, the behaviorally different notions of dominance solvability and global stability coincide. This is similar to the case for strategic complements.

A difference is that uniqueness of Nash equilibrium is not necessarily sufficient for this conclusion, as shown by the three firm Cournot oligopoly in Example 4.33 (page 112). Even though $\underline{x}$ and $\overline{x}$ are profiles of simply rationalizable strategies, it is possible that neither $\underline{x}$ nor $\overline{x}$ is a Nash equilibrium.

Another difference is that in GSS, best response dynamic from $\inf X$ (or from $\sup X$) may not necessarily converge, but if one of these converges, then the other one converges as well, and both converge to the same limit. In GSC, both best response dynamics always converge and it is not necessarily true that they converge to the same limit.

An implication is that in GSS, a single best response dynamic has important information about global stability and dominance solvability. Whether best response dynamic from inf X or from sup X is a good model for behavior in a game or not, convergence of either implies convergence of every adaptive dynamic to the same limit and a unique prediction, and equivalently, a unique prediction using serially undominated strategies.

The theorem requires only that the best response dynamic converges, and does not require computation of the limit. This is a convenient tool to check for robustness of equilibrium in GSS.

This theorem combined with totally unordered Nash equilibrium set yields the following conditions to determine uniqueness of Nash equilibrium in GSS.

Theorem 4.29 *In a GSS, suppose the Nash equilibrium set is totally unordered. Let $\underline{x}$ be the limit of the lower mixture of best response dynamics from* inf X *and* sup X, *and $\overline{x}$ be the limit of the upper mixture.*

1. *$\underline{x}$ is a Nash equilibrium, if, and only if, $\underline{x}$ is the unique Nash equilibrium in the game.*
2. *$\overline{x}$ is a Nash equilibrium, if, and only if, $\overline{x}$ is the unique Nash equilibrium in the game.*
3. *With multiple Nash equilibria, neither $\underline{x}$ nor $\overline{x}$ are Nash equilibria.*

Proof For statement (1), only sufficiency needs to be proved. If the game has a Nash equilibrium x different from $\underline{x}$, then Nash equilibrium is serially undominated implies $x \in [\underline{x}, \overline{x}]$, and therefore, $\underline{x} \preceq x$, contradicting the supposition that Nash equilibrium set is totally unordered. Statement (2) follows similarly. Statement (3) follows from statements (1) and (2). □

This theorem provides another contrast with GSC (Theorem 3.23, page 69) where limits of the corresponding bounds on the set of serially undominated strategies are both Nash equilibria and may easily be different.

This theorem provides computational algorithms to check for uniqueness of Nash equilibrium. The algorithms given by the lower mixture and the upper mixture of best response dynamics are both convergent algorithms. In finite games, each converges in finitely many iterations. Checking if the limit is a Nash equilibrium gives an algorithmic test for uniqueness.

Consider the following examples with affine best responses from Roy and Sabarwal (2012). Their stability analysis is different here. Roy and Sabarwal (2012) consider the second iterate of joint best response. The analysis here uses the best response itself (not its second iterate). This gives more direct results.

Example 4.30 (Affine best responses) Several applications of canonical models have affine best response functions. A general class of such models is the following. Suppose $\Gamma = (X_i, u_i)_{i=1}^{I}$ is a GSS in which for every i, $X_i \subset [0, \overline{x}_i] \subset \mathbb{R}$. Suppose best response of each player i is singleton valued, $B_i : X_{-i} \to X_i$, and $B : X \to X$ is the joint best response function.

Suppose B has the form $B(x) = a + Ax$, where $a \in \mathbb{R}^I$ and A is a $I \times I$ matrix with nonpositive entries. It follows that B is a (weakly) decreasing function on X. For range of B to be in X, let $-A\overline{x} \leq a \leq \overline{x}$.

Let $(y^n)_{n=0}^{\infty}$ be best response dynamic from inf X, with $y^0 = \inf X$, and for $n \geq 0$, $y^n = B(y^{n-1})$. In this case,

$$\begin{aligned} \|B(y^n) - B(y^{n-1})\| &= \|A(y^n - y^{n-1})\| \\ &\leq \|A\|_\infty \|y^n - y^{n-1}\|_\infty, \end{aligned}$$

where $\|x\|_\infty = \max_i |x_i|$ is the maximum of absolute value of components of x, and $\|A\|_\infty = \max_i\{\sum_j |a_{i,j}|\}$ is maximum of sum of absolute value of row entries of A. If $\|A\|_\infty < 1$, then best response is a contractive mapping and best response dynamic from inf X converges.

Moreover, the game has a unique Nash equilibrium, if, and only if, there is unique x^* satisfying $x^* = a + Ax^*$, and this is equivalent to invertibility of $(I - A)$. In this case, $x^* = (I - A)^{-1}a$. Some specific examples follow.

Example 4.31 (Shared output games) Suppose there is a group of three persons, each with an amount to invest in two projects. One project is a team project that yields an output with value based on total investment of all players. This value is shared proportionally based on investment of each person. The other project yields a private return. Both projects have diminishing marginal product.

Suppose total investment amount available to each person is $w > 0$. Each person chooses an amount $x_i \in [0, w]$ to invest in the team project and the remainder $w - x_i$ to invest in the private project.

Given investment of each person, value of output of team project is given by a concave technology, $a(x_1 + x_2 + x_3) - b(x_1 + x_2 + x_3)^2$, and person i receives a proportionate share $\frac{x_i}{x_1+x_2+x_3}(a(x_1+x_2+x_3) - b(x_1+x_2+x_3)^2)$. The private project yields $\alpha(w-x_i) - \beta(w-x_i)^2$ to person i. Here, $a > \alpha > 0$ and $b > \beta > 0$. Payoff to person i is $u_i(x_1, x_2, x_3) = \frac{x_i}{x_1+x_2+x_3}(a(x_1 + x_2 +$

$x_3) - b(x_1 + x_2 + x_3)^2) + \alpha(w - x_i) - \beta(w - x_i)^2$. It is easy to check that this payoff has decreasing differences in (x_i, x_{-i}).

Best response of person i is $B_i(x_j, x_k) = \frac{a-\alpha+2\beta w}{2b+2\beta} - \frac{b}{2b+2\beta}(x_j + x_k)$. For range of best response to be in $[0, w]$, assume that $\frac{a-\alpha}{2b} \leq w \leq \frac{a-\alpha}{2(b-\beta)}$. This is an affine best response and the corresponding matrix A is given by

$$A = \frac{1}{2b + 2\beta} \begin{pmatrix} 0 & -b & -b \\ -b & 0 & -b \\ -b & -b & 0 \end{pmatrix},$$

with $\|A\|_\infty = \frac{2b}{2b+2\beta} < 1$. It follows that this game is dominance solvable, is globally stable, and has a unique Nash equilibrium.

Example 4.32 (Cournot oligopoly) Suppose there are I firms in an industry with differentiated products, linear demand, and linear cost. Inverse demand for output of firm i is given by $p_i = \alpha - \beta x_i - \delta(\sum_{j \neq i} x_j)$, where x_i is output of firm i and $\sum_{j \neq i} x_j$ is sum of outputs of firms other than i. Marginal cost for each firm is the same, and constant at $c > 0$. Each firm can produce output subject to a production capacity constraint given by $x_i \in [0, x^{max}]$. Profit of firm i is given by

$$u_i(x_1, \ldots, x_I) = \left(\alpha - \beta x_i - \delta \left(\sum_{j \neq i} x_j \right) \right) x_i - c x_i.$$

Maximization with respect to x_i yields

$$B_i(x_{-i}) = \frac{\alpha - c}{2\beta} - \frac{\delta}{2\beta} \left(\sum_{j \neq i} x_j \right).$$

For range of best response to be in $[0, x^{max}]$, assume that $\frac{\alpha-c}{2\beta} < x^{\max}$ (the strict inequality ensures that best response does not hit maximum output and this ensures stability for all values), and $\delta(I - 1)x^{max} \leq \alpha - c$. Best response of each firm is affine and the corresponding matrix A is given by

$$A = \frac{\delta}{2\beta} \begin{pmatrix} 0 & -1 & \cdots & -1 \\ -1 & 0 & \cdots & -1 \\ \vdots & \vdots & \ddots & \vdots \\ -1 & -1 & \cdots & 0 \end{pmatrix}.$$

In this case, $\|A\|_\infty = \frac{(I-1)\delta}{2\beta} = \frac{(I-1)\delta}{2\beta} \cdot \frac{x^{max}}{x^{max}} \le \frac{\alpha-c}{2\beta x^{max}} < 1$. Therefore, this Cournot oligopoly is dominance solvable, is globally stable, and has a unique Nash equilibrium.

Example 4.33 (Cournot oligopoly-2) In the Cournot oligopoly above (Example 4.32, page 111), the condition $\delta(I-1)x^{max} \le \alpha - c < 2\beta x^{\max}$ implies $I - 1 \le \frac{\alpha-c}{\delta x^{max}} < \frac{2\beta}{\delta}$. This gives an upper bound on the number of competitors $(I-1)$ in terms on heterogeneity parameters β and δ. The parameter β measures effect of a firm's output on its own price and δ measures effect of competitor outputs on own price. For example, if own output effect is large, say, $\beta = 10\delta$, then global stability holds in an industry with $I \le 20$ firms. On the other hand, if output is homogeneous, with $\beta = \delta$, then global stability holds only in a duopoly. With $\beta = \delta$, if there are three firms, best response dynamic from $\inf X$ cycles, $(0,0,0) \mapsto (\frac{\alpha-c}{2\beta}, \frac{\alpha-c}{2\beta}, \frac{\alpha-c}{2\beta}) \mapsto (0,0,0)$. This result may be avoided with more general demand and cost curves.

A new example with nonlinear best responses is the following.

Example 4.34 (Worker competition) Consider a firm in which three workers are competing for a reward. The worth of this reward for each worker is normalized to 1. Each worker chooses intensity of effort $x_i \in [0, 1]$. There is an expected benchmark effort in the firm, normalized to $\frac{1}{2}$. Cost of effort to worker i from effort x_i is given by quadratic deviation from benchmark $(x_i - \frac{1}{2})^2$. Intuitively, working higher than benchmark is costly because of additional effort and working lower than benchmark is costly because of negative performance reviews. Moreover, collective intensity of effort reduces reward by $x_1x_2x_3$ due to competition. In other words, payoff for worker i is given by

$$u_i(x_1, x_2, x_3) = 1 - x_1x_2x_3 - \left(x_i - \frac{1}{2}\right)^2.$$

In this case, best response of worker i is given by

$$B_i(x_j, x_k) = \frac{1}{2} - \frac{1}{2}x_jx_k.$$

This is a nonlinear and (weakly) decreasing function of (x_j, x_k). It is strictly decreasing when $x_j > 0$ and $x_k > 0$. The best response dynamics from

(0, 0, 0) is given by $(0,0,0) \mapsto (\frac{1}{2}, \frac{1}{2}, \frac{1}{2}) \mapsto (\frac{3}{8}, \frac{3}{8}, \frac{3}{8}) \mapsto (\frac{55}{128}, \frac{55}{128}, \frac{55}{128})$, and so on. It can be checked that this is a (nonmonotone) convergent sequence. Therefore, this game has a unique Nash equilibrium, is dominance solvable, and globally stable. For reference, the unique Nash equilibrium is $(\sqrt{2}-1, \sqrt{2}-1, \sqrt{2}-1) \approx (0.42, 0.42, 0.42)$. This is below benchmark due to strategic substitutes effect of competition.

4.2.5 *Monotone Comparative Statics of Equilibrium*

For situations with contradirectional incentives, the effect of a change in the decision-making environment on predicted outcomes is studied using parameterized games with strategic substitutes. These are a subclass of parameterized lattice games.

In a parameterized lattice game $\Gamma = ((X_i, u_i)_{i=1}^I, T)$, for each $t \in T$, ***player i has strategic substitutes at*** t, if the section of player i payoff determined by t, $u_i(x_i, x_{-i}, t)$, has dual single crossing property in (x_i, x_{-i}) and is quasisupermodular on X_i (for each x_{-i}).

As motivated in Sects. 2.2.5 (page 41) and 3.2.5 (page 73), in situations where changes in the decision-making environment are important, the question of interest may be formulated in the following manner. If decision-making environment is made more favorable to take a higher action, when will equilibrium actions go up? In games with strategic substitutes, this is not always the case, as shown by the parameterized asymmetric Cournot duopoly (Example 4.38, page 119).

The idea that an increase in parameter t provides to player i an incentive to take a higher action is formalized as follows. ***Player i has incentive to take higher action in*** t, if $u_i(x_i, x_{-i}, t)$ has single crossing property in (x_i, t), for every x_{-i}.

A ***parameterized game with strategic substitutes***, or ***parameterized GSS***, is a parameterized lattice game $\Gamma = ((X_i, u_i)_{i=1}^I, T)$, where

1. There are finitely many players, I, indexed by $i \in \{1, \ldots, I\}$.
2. For each i, the action space of player i is X_i. It is assumed to be a nonempty, subcomplete lattice in $\mathbb{R}^{n_i}$ with the Euclidean order. The space of profiles of actions is $X = X_1 \times \cdots \times X_I$ with the product order. The parameter space is $T \subset \mathbb{R}^n$. It is a poset with the Euclidean partial order.

3. For each i, the payoff for player i is $u_i : X_i \times X_{-i} \times T \to \mathbb{R}$, where

    (a) u_i is upper semicontinuous on X and continuous on X_{-i}, for every t,
    (b) u_i has dual single crossing property in (x_i, x_{-i}) (for each t), u_i is quasisupermodular on X_i (for each (x_{-i}, t)), and
    (c) u_i has single crossing property in (x_i, t) (for each x_{-i}).

In other words, a parameterized GSS is a parameterized lattice game in which every player i has strategic substitutes at every t and every player i has incentive to take a higher action in t.

It follows immediately from the definition that in a parameterized GSS, for every $t \in T$, the game at t, $\Gamma(t)$, is a GSS. In particular, the section of best response correspondence for player i determined by t, $B_i(\cdot, t)$, mapping (x_{-i}, t) to $B_i(x_{-i}, t)$ is nonempty valued, compact valued, subcomplete valued, upper hemicontinuous in x_{-i}, and (weakly) decreasing in x_{-i}.

Moreover, single crossing property in (x_i, t) implies that for each x_{-i}, the section of best response correspondence for player i determined by x_{-i}, $B_i(x_{-i}, \cdot)$, mapping (x_{-i}, t) to $B_i(x_{-i}, t)$ is (weakly) increasing in t.

Similarly, for each t, the section of joint best response correspondence determined by t, $B(\cdot, t)$, mapping (x, t) to $B(x, t)$ is nonempty valued, compact valued, subcomplete valued, upper hemicontinuous in x, and (weakly) decreasing in x, and moreover, for each x, the section of joint best response correspondence determined by x, $B(x, \cdot)$ mapping (x, t) to $B(x, t)$ is (weakly) increasing in t.

The equilibrium correspondence is $\mathcal{E} : T \rightrightarrows X$, mapping t to $\mathcal{E}(t)$.

Differences in patterns of equilibrium behavior in GSS and GSC manifest themselves in different properties for their equilibrium correspondences.

In GSC, the equilibrium set is a nonempty, complete lattice. This identifies a smallest and largest equilibrium and monotone response of these equilibria to the parameter implies that every parameterized GSC has monotone comparative statics of extremal equilibrium outcomes.

More generally, in a lattice game, the equilibrium set is totally unordered under mild conditions on one or two players only (Theorem 4.21, page 101), and therefore, with multiple equilibria, it is impossible to identify a smallest equilibrium or a largest equilibrium. Consequently, monotone comparative statics of extremal equilibrium outcomes is an impossibility in such situations.

Different techniques are needed to show monotone comparative statics of equilibrium outcomes in these cases. As a step in this direction, the next theorem shows that under mild conditions, equilibrium in a parameterized lattice game never goes down, that is, the equilibrium correspondence in a parameterized lattice game is never decreasing. Recall that $\mathcal{E} : T \rightrightarrows X$ is never decreasing, if for every $\hat{t} \prec \tilde{t}$, for every $x^* \in \mathcal{E}(\hat{t})$, and for every $y^* \in \mathcal{E}(\tilde{t})$, $y^* \not\prec x^*$.

Recall that for subsets A, B of poset X, A is completely lower than B, denoted $A \sqsubset_c B$, if for every $a \in A$ and for every $b \in B$, $a \prec b$. In a parameterized lattice game, ***player i has strict strategic substitutes at t***, if for every $\hat{x}_{-i}$ and $\tilde{x}_{-i}$ in X_{-i}, $\hat{x}_{-i} \prec \tilde{x}_{-i}$ implies $B_i(\tilde{x}_{-i}, t) \sqsubset_c B_i(\hat{x}_{-i}, t)$. In other words, player i has strict strategic substitutes if $B_i(\cdot, t)$ is decreasing in the completely lower than set order.

It is useful to have a weaker notion as well. For subsets A, B of poset X, ***A is uniformly lower than B*** , denoted $A \sqsubseteq_u B$, if for every $a \in A$ and for every $b \in B$, $a \preceq b$. For $\hat{t}, \tilde{t} \in T$ with $\hat{t} \prec \tilde{t}$, ***player i has uniformly higher actions at*** $(\hat{t}, \tilde{t})$, if for every x_{-i}, $B_i(x_{-i}, \hat{t}) \sqsubseteq_u B_i(x_{-i}, \tilde{t})$. In other words, player i has uniformly higher actions at $\hat{t} \prec \tilde{t}$, if when t increases from $\hat{t}$ to $\tilde{t}$, $B_i(x_{-i}, \hat{t})$ increases to $B_i(x_{-i}, \tilde{t})$ in the uniformly lower than set order.

Theorem 4.35 (Never decreasing equilibrium) *In every parameterized lattice game in which every player has incentive to take higher action in t,*

1. *If for every $\hat{t} \prec \tilde{t}$, there is player i such that player i has strict strategic substitutes and singleton valued best response at $t \in \{\hat{t}, \tilde{t}\}$, or*
2. *If for every $\hat{t} \prec \tilde{t}$, there is player i such that player i has strict strategic substitutes at $t \in \{\hat{t}, \tilde{t}\}$ and uniformly higher actions at $(\hat{t}, \tilde{t})$, or*
3. *If for every t, there are two players with strict strategic substitutes at t,*

Then the equilibrium correspondence is never decreasing.

Proof Consider $\hat{t} \prec \tilde{t}$, $x^* \in \mathcal{E}(\hat{t})$, and $y^* \in \mathcal{E}(\tilde{t})$.

Suppose condition (1) is satisfied, without loss of generality, for player 1. If $y^* \prec x^*$, then either $y^*_{-1} \prec x^*_{-1}$, or $y^*_{-1} = x^*_{-1}$ and $y^*_1 \prec x^*_1$. If $y^*_{-1} \prec x^*_{-1}$, then singleton best response, strict strategic substitutes, and single crossing property in (x_1, t) imply that $\{x^*_1\} = B_1(x^*_{-1}, \hat{t}) \sqsubset_c B_1(y^*_{-1}, \hat{t}) \sqsubseteq B_1(y^*_{-1}, \tilde{t}) = \{y^*_1\}$. Therefore, $x^*_1 \prec y^*_1$, and this contradicts

$y^* \prec x^*$. If $y^*_{-1} = x^*_{-1}$ and $y^*_1 \prec x^*_1$, then $\{x^*_1\} = B_1(x^*_{-1}, \hat{t}) = B_1(y^*_{-1}, \hat{t}) \sqsubseteq B_1(y^*_{-1}, \tilde{t}) = \{y^*_1\}$, and this contradicts $y^*_1 \prec x^*_1$.

Suppose condition (2) is satisfied, without loss of generality, for player 1. If $y^* \prec x^*$, then either $y^*_{-1} \prec x^*_{-1}$, or $y^*_{-1} = x^*_{-1}$ and $y^*_1 \prec x^*_1$. If $y^*_{-1} \prec x^*_{-1}$, then strict strategic substitutes and uniformly higher actions imply $B_1(x^*_{-1}, \hat{t}) \sqsubset_c B_1(y^*_{-1}, \hat{t}) \sqsubseteq_u B_1(y^*_{-1}, \tilde{t})$, and therefore, $x^*_1 \preceq \overline{B}_1(x^*_{-1}, \hat{t}) \prec \underline{B}_1(y^*_{-1}, \hat{t}) \preceq \underline{B}_1(y^*_{-1}, \tilde{t}) \preceq y^*_1$, contradicting $y^* \prec x^*$. If $y^*_{-1} = x^*_{-1}$ and $y^*_1 \prec x^*_1$, then uniformly higher actions imply $B_1(x^*_{-1}, \hat{t}) = B_1(y^*_{-1}, \hat{t}) \sqsubseteq_u B_1(y^*_{-1}, \tilde{t})$, and therefore, $x^*_1 \preceq \overline{B}_1(x^*_{-1}, \hat{t}) = \overline{B}_1(y^*_{-1}, \hat{t}) \preceq \underline{B}_1(y^*_{-1}, \tilde{t}) \preceq y^*_1$, contradicting $y^*_1 \prec x^*_1$.

If condition (3) is satisfied, then proof of statement (2) in Theorem 4.21 (page 101) shows that for each t, $B(x, t)$ is never increasing in x. Consider $\hat{t} \prec \tilde{t}$, $x^* \in \mathcal{E}(\hat{t})$, $y^* \in \mathcal{E}(\tilde{t})$, and suppose $y^* \prec x^*$. Let $A = \{x \in X \mid \exists x' \in B(x, \tilde{t}), x \preceq x'\}$. Using the same argument as in proof of Theorem 2.21 (page 28), it follows that y^* is a maximal element of A. Moreover, $B(x, t)$ is (weakly) increasing in t implies $B(x^*, \hat{t}) \sqsubseteq B(x^*, \tilde{t})$, and therefore, $x^* \preceq \overline{B}(x^*, \tilde{t})$. As $\overline{B}(x^*, \tilde{t}) \in B(x^*, \tilde{t})$, it follows that $x^* \in A$, and this contradicts y^* is a maximal element of A. □

In conditions (1) of Theorem 4.35 (page 115), the requirement is for one player to have strictly decreasing best response function at each pair of ordered parameter values. The identity of the player may change for different pairs of parameter values. This allows for cases where a player has strategic substitutes for some parameter values and not for other parameter values. In particular, it allows for the case when no player has strategic substitutes at all parameter values. This generalizes the result due to Monaco and Sabarwal (2016) for the special case where one player has strict strategic substitutes at all parameter values.

Conditions (2) and (3) are helpful if best response is not singleton valued.

In condition (2), the requirement is for one player to have strictly decreasing best response correspondence at each pair of ordered parameter values and incentive to take uniformly higher action at those parameter values. As in condition (1), the identity of the player may change for different pairs of parameter values. This allows for cases where a player satisfies these conditions for some parameter values and not for other parameter values. In particular, it allows for the case when no player satisfies these conditions for all parameter values. This generalizes the result due to Monaco and Sabarwal (2016) for the special case where one player satisfies these conditions for all parameter values.

Condition (3) is a new condition unavailable in the previous literature. The requirement is for two players to have strict strategic substitutes at one parameter value rather than at two parameter values. The identity of these two players may change with the parameter value. It also allows for cases when no player has strategic substitutes at all parameter values. Moreover, it does not require uniformly higher actions in t. This is helpful when best responses are correspondences and players have incentive for higher action in t but not uniformly higher action in t.

In Theorem 4.35 (page 115), all three conditions refer to one or two players only. As long as these conditions are satisfied, there are no assumptions about payoffs for other players at other parameter values. Other players may have strategic complements, or strategic substitutes, or neither, and in each case, the equilibrium correspondence is never decreasing.

As a special case, Theorem 4.35 (page 115) shows that under its conditions, the equilibrium correspondence in every parameterized GSS is never decreasing.

The equilibrium correspondence in parameterized GSC does not necessarily satisfy this result, because in those games, it is easy to have an equilibrium at a higher parameter value (say, the smallest one) that is lower than an equilibrium at a lower parameter value (say, the highest one), even with symmetric equilibrium, as shown by discrete Bertrand duopoly (Example 3.28, page 75).

Theorem 4.35 (page 115) implies that with a minimal change to a GSC, such as addition on one player with strictly decreasing best response function, decreasing selection of equilibrium is ruled out completely. This is another implication of moving away from GSC framework. When action space of each player is a chain, this distinction is seen even more clearly.

Theorem 4.36 (MCS-Chains) *In every parameterized lattice game in which action space of each player is a chain, if the equilibrium correspondence is never decreasing, then*

1. *For every* $\hat{t} \prec \tilde{t}$, *for every* $x^* \in \mathcal{E}(\hat{t})$, *and for every* $y^* \in \mathcal{E}(\tilde{t})$, *there is player* i *such that* $x_i^* \preceq y_i^*$, *and*
2. *For every* $\hat{t} \prec \tilde{t}$, *for every symmetric equilibrium* $x^* \in \mathcal{E}(\hat{t})$, *and for every symmetric equilibrium* $y^* \in \mathcal{E}(\tilde{t})$, $x^* \preceq y^*$.

Proof For statement (1), equilibrium correspondence is never decreasing implies that for every $\hat{t} \prec \tilde{t}$, for every $x^* \in \mathcal{E}(\hat{t})$, and for every $y^* \in \mathcal{E}(\tilde{t})$, $y^* \not\prec x^*$. If for every player i, $x_i^* \not\preceq y_i^*$, then as strategy space of each player is a chain, it follows that for every player i, $y_i^* \prec x_i^*$, and therefore, $y^* \prec x^*$, a contradiction. Statement (2) follows from statement (1), because in a symmetric equilibrium, each player plays the same action, and therefore, if there is player i with $x_i^* \preceq y_i^*$, then for every i, $x_i^* \preceq y_i^*$, and it follows that $x^* \preceq y^*$. □

Under statement (1) in this theorem, when the parameter goes up, regardless of the equilibrium considered at the lower parameter value, and regardless of the equilibrium considered at the higher parameter value, there is always at least one player whose action in the new equilibrium is higher than it was in the old equilibrium.

In a symmetric equilibrium, every player plays the same action, and therefore, statement (2) in this theorem shows that every symmetric equilibrium at a higher parameter value is higher than every symmetric equilibrium at a lower parameter value. In other words, every selection of symmetric equilibria is (weakly) increasing.

Both conclusions are not necessarily true for a parameterized GSC, as shown by the discrete Bertrand duopoly (Example 3.28, page 75).

Conditions that guarantee never decreasing correspondence may also guarantee unique symmetric equilibrium, and therefore, these may be combined to conclude that symmetric equilibrium selection is uniquely defined and (weakly) increasing.

Theorem 4.37 (MCS-Symmetric equilibrium) *In a parameterized lattice game in which each player has incentive to take higher action in t and in which action space of each player is a chain, if*

1. *For every $\hat{t} \prec \tilde{t}$, there is player i such that player i has strict strategic substitutes and singleton valued best response at $t \in \{\hat{t}, \tilde{t}\}$, or*
2. *For every t, there are two players with strict strategic substitutes at t,*

Then on the subset of parameters for which the symmetric Nash equilibrium set is nonempty,

1. *For every t, the symmetric Nash equilibrium at t is unique, denoted $x^*(t)$, and*
2. *The mapping $t \mapsto x^*(t)$ is (weakly) increasing.*

Proof Condition (1) or (2) combined with Theorem 4.21 (page 101) implies that for every t, the Nash equilibrium set is totally unordered. Therefore, Corollary 4.24 (page 104) implies that on the subset of parameters for which the symmetric Nash equilibrium set is nonempty, for every t, the symmetric Nash equilibrium at t is unique, denoted $x^*(t)$.

Moreover, conditions (1) and (2) combined with Theorem 4.35 (page 115) imply that the equilibrium correspondence is never decreasing, and therefore, statement (2) in Theorem 4.36 (page 117) implies that on the subset of parameters for which the symmetric Nash equilibrium set is nonempty, the mapping $t \mapsto x^*(t)$ is (weakly) increasing. □

The conclusion in this theorem is not necessarily true for parameterized GSC, as shown by discrete Bertrand duopoly (Example 3.28, page 75), which has multiple symmetric Nash equilibria, and a strictly lower symmetric equilibrium at a higher parameter value than one at a lower parameter value.

Like the previous two theorems, Theorem 4.37 (page 118) is formulated at the more general level of parameterized lattice games. It imposes requirements for one or two players only, and no assumptions about payoffs for other players. Other players may have strategic complements, or strategic substitutes, or neither, and in each case, the symmetric equilibrium correspondence is (weakly) increasing.

More generally, nonsymmetric Nash equilibrium may not necessarily be (weakly) increasing. Consider the following example due to Roy and Sabarwal (2010).

Example 4.38 (Parameterized asymmetric Cournot duopoly) Consider two firms producing an output with a positive societal impact, say, wifi hotspots, and competing as Cournot duopoly. In the absence of market intervention, total output is lower than socially optimal. In order to increase output, a policy maker subsidizes cost of production for each firm. In equilibrium, will a firm produce more output if it receives a subsidy?

Suppose inverse market demand is given by $p = a - b(x_1 + x_2)$, where x_1 is output of firm 1 and x_2 is output of firm 2, and each firm has constant marginal cost of production given by $c > 0$. A policy maker provides a subsidy of $t > 0$ per unit to lower the cost of production and allocates share $\alpha \in [0, 1]$ of the subsidy to firm 1 and share $(1 - \alpha)$ to firm 2. After the subsidy, marginal cost of firm 1 is $c_1 = c - \alpha t$ and that of firm 2 is $c_2 = c - (1 - \alpha)t$. In equilibrium, will a firm produce more output if subsidy t increases?

Another interpretation is that t represents technological innovation (leading to lower marginal cost) and α measures heterogeneity in adoption of technological innovation across firms. In equilibrium, will a firm produce more output if technological innovation t goes up?

Firm profit is given by

$$u_1(x_1, x_2) = (a - b(x_1 + x_2))x_1 - (c - \alpha t)x_1$$
$$u_2(x_1, x_2) = (a - b(x_1 + x_2))x_2 - (c - (1 - \alpha)t)x_2.$$

Best response of firm 1 is $B_1(x_2) = \frac{a-c+\alpha t}{2b} - \frac{1}{2}x_2$ and best response of firm 2 is $B_2(x_1) = \frac{a-c+(1-\alpha)t}{2b} - \frac{1}{2}x_1$. The unique equilibrium output is

$$x_1^*(t) = \frac{a - c + (3\alpha - 1)t}{3b} \quad \text{and} \quad x_2^*(t) = \frac{a - c + (2 - 3\alpha)t}{3b}.$$

It follows that

$$0 < \alpha < \tfrac{1}{3} \Leftrightarrow x_1^*(t) \text{ is decreasing in } t \text{ and } x_2^*(t) \text{ is increasing in } t,$$
$$\tfrac{1}{3} < \alpha < \tfrac{2}{3} \Leftrightarrow x_1^*(t) \text{ is increasing in } t \text{ and } x_2^*(t) \text{ is increasing in } t, \text{ and}$$
$$\tfrac{2}{3} < \alpha < 1 \Leftrightarrow x_1^*(t) \text{ is increasing in } t \text{ and } x_2^*(t) \text{ is decreasing in } t.$$

In the example, when the parameter is interpreted as a subsidy, the implication is that for output of both firms to increase, the subsidy cannot be too heterogeneous. If firm 1 gets less than one-third of the subsidy its output goes down, and firm 2 becomes larger and more dominant. If firm 1 gets more than two-thirds of the subsidy, firm 2 output goes down and firm 1 becomes larger and more dominant. In the intermediate case, output of both firms goes up.

When the parameter is interpreted as technological innovation, the implication is that to prevent imbalances in firm size, technological adoption should not be too asymmetric. Otherwise, there is greater chance of market dominance by one firm at the expense of the other, and a corresponding impact on market power.

Additional analysis of this example shows the following.

For the theory of monotone comparative statics of equilibrium outcomes, this example shows that even with two players, each with action space that is a chain, each with best response that is a strictly decreasing function, and an equilibrium correspondence that is never decreasing, in the absence of symmetry in equilibrium, it does not necessarily follow that every player takes a higher equilibrium action. In other words, in general, statement (1) of Theorem 4.36 (page 117) cannot be strengthened to conclude that every player takes a higher action. Consequently, even in otherwise well-behaved cases, there may be no monotone comparative statics of equilibrium outcomes.

On the other hand, in this example, total industry output in equilibrium is

$$x_1^*(t) + x_2^*(t) = \frac{2(a-c)+t}{3b}.$$

This is increasing in t and does not depend on α. This has two implications. First, if only the sum of actions is a variable of interest, then that can increase even when all the addends do not. In this case, some addends may go up and others may go down. Second, if it is important that no player's action goes down, then we are back in the theory of monotone comparative statics for the full profile of equilibrium actions. Moreover, if no player's action goes down and some player's action goes up, then the sum goes up automatically. This version results in both sum of actions going up and no addend going down.

A new analysis of Example 4.38 (page 119) highlights a limitation of the attempt to change a two player GSS into a GSC by reversing the order on actions of one player. Reversing the order on the quantity of firm one by defining $y_1 = -x_1$, using y_1 as the decision variable for firm 1, and replacing x_1 with $-y_1$ in both payoffs makes this a GSC, because $\frac{\partial^2 u_1}{\partial x_2 \partial y_1} = \frac{\partial^2 u_2}{\partial y_1 \partial x_2} = b > 0$. But this implies $\frac{\partial^2 u_1}{\partial t \partial y_1} = -\alpha < 0$, violating the single crossing property in (y_1, t). Going one step further, let $s = -t$, use s as the parameter, and replace t with $-s$ in the payoff for both players. Then $\frac{\partial^2 u_1}{\partial s \partial y_1} = \alpha > 0$, but $\frac{\partial^2 u_2}{\partial s \partial x_2} = -(1-\alpha) < 0$, which violates single crossing property in (x_2, s). Trying to fix this by replacing t with $-s$ for firm 1 and t with $s = -t$ for firm 2 changes the problem materially by reversing the interpretation of the subsidy for firm 2. In this case, an increase in s implies an increase in best choice of x_2, but an increase in s is the same as a decrease in t. In other words, lowering the subsidy t (an increase in s) causes output of firm 1 to go down (increase in y_1) but causes output of firm 2 to go up (increase in x_2), thereby reversing the nature of the subsidy for firm 2 in the original example. A similar analysis can be performed starting with a reversal of the order on quantity of firm 2.

Example 4.38 (page 119) cannot be transformed into a parameterized GSC by reversing either the order on the action of one player or the order on the parameter without altering the example materially. Indeed, as the solution shows, regardless of which firm's order is reversed, there are ranges of α for which there is no monotone comparative statics of equilibrium in the original variables even though the equilibrium is unique and action spaces are chains.

In parameterized GSS, monotone comparative statics of equilibrium outcomes may fail because the direct and indirect effects of an increase in the parameter are working against each other. When a parameter goes up, single crossing property in (x_i, t) gives each player an incentive to take a higher action. This is the direct effect of parameter increase on a player's decision. When competitors take higher actions due to an increase in the parameter, dual single crossing property in (x_i, x_{-i}) gives each player an incentive to take a lower action. This is the indirect strategic substitutes effect of parameter increase and it is going in the direction opposite to the direct effect. The final outcome depends on the tradeoff between these opposing effects.

In parameterized GSC, both effects work in the same direction. The direct parameter effect is the same as above and provides an incentive to take a higher action. When competitors take higher actions due to an increase in the parameter, single crossing property in (x_i, x_{-i}) gives each player an incentive to increase their action further. The indirect strategic complements effect reinforces the direct parameter effect and the combined effect leads to a higher final outcome for every player.

In Example 4.38 (page 119) above, the heterogeneity parameter α may be viewed as intensity of the direct effect for player 1 and $(1-\alpha)$ may be viewed as intensity of the direct effect for player 2. When α is large relative to $(1-\alpha)$, the direct effect is large for player 1 and overpowers the indirect strategic substitutes effect leading to a higher outcome for player 1. For player 2, $(1-\alpha)$ is low relative to α, and therefore, the direct parameter effect is small for player 2 and cannot overpower the indirect strategic substitutes effect leading to a lower outcome for player 2. The conclusion is reversed when α is small relative to $(1-\alpha)$.

In parameterized GSS, sufficient conditions for monotone comparative statics of profile of equilibrium actions may be formulated in terms of a tradeoff between direct and indirect effects of a parameter increase.

For two player case, the appropriate tradeoff between direct and indirect effects is formulated as follows. Consider a parameterized GSS with two players, an equilibrium x^* at parameter $\hat{t}$, and suppose parameter increases to $\tilde{t}$.

For convenience, suppose best response of both players is singleton valued. For each player, the direct effect is summarized by the best response at the new parameter, holding fixed the other player's equilibrium action, that is, by $B_i(x^*_{-i}, \tilde{t})$. The indirect effect is summarized by the best response to the best response of the other player due to the direct effect, holding fixed the parameter for the given player, that is, by $B_i(B_j(x^*_{-j}, \tilde{t}), \hat{t})$. The combined effect is given by the composition of the two effects, that is, by $B_i(B_j(x^*_{-j}, \tilde{t}), \tilde{t})$. If for each player, this combined effect is favorable, in the sense that the combined effect is a sufficient incentive for the player to play an action higher than the old equilibrium action, that is, $x^*_i \preceq B_i(B_j(x^*_{-j}, \tilde{t}), \tilde{t})$, then there is an equilibrium at $\tilde{t}$ that is higher than x^*.

When best response is a correspondence, the directionally extremal effects do the trick. That is, if $x^*_i \preceq \underline{B}_i(\overline{B}_j(x^*_{-j}, \tilde{t}), \tilde{t})$ for both players, then there is an equilibrium at $\tilde{t}$ that is higher than x^*. Here, $\overline{B}_j(x^*_{-j}, \tilde{t}) = \sup B_j(x^*_{-j}, \tilde{t})$ and $\underline{B}_i(\overline{B}_j(x^*_{-j}, \tilde{t}), \tilde{t}) = \inf B_i(\overline{B}_j(x^*_{-j}, \tilde{t}), \tilde{t})$.

This intuition generalizes to more than two players.

The next two theorems are new results unavailable in the previous literature. They do not invoke a fixed point theorem, or convex action spaces, or convex valued best responses. They present conditions under which, for a general parameterized GSS, monotone comparative statics of equilibrium outcomes is obtained.

With more than two players, it is useful to have notation for joint direct parameter effect. Consider a parameterized GSS, an equilibrium x^* at parameter $\hat{t}$, and suppose parameter increases to $\tilde{t}$. The joint direct parameter effect is $\tilde{y} = \overline{B}(x^*, \tilde{t})$, where $\overline{B}(x^*, \tilde{t}) = \sup B(x^*, \tilde{t})$ is supremum of the joint best response set at $(x^*, \tilde{t})$.

The joint composite effect is given by $\underline{B}(\tilde{y}, \tilde{t}) = \inf B(\tilde{y}, \tilde{t})$, and as shown next, if $x^* \preceq \underline{B}(\tilde{y}, \tilde{t})$, then a higher profile of equilibrium actions may be guaranteed at $\tilde{t}$.

Theorem 4.39 (MCS-Parameterized GSS) *In a parameterized GSS, for each* $\hat{t}, \tilde{t} \in T$ *with* $\hat{t} \prec \tilde{t}$, *and for each* $x^* \in \mathcal{E}(\hat{t})$, *let* $\tilde{y} = \overline{B}(x^*, \tilde{t})$, *let* $\tilde{\Gamma}(\tilde{t})$ *be the lattice game at* $\tilde{t}$ *restricted to* $[x^*, \tilde{y}]$, *and let* $\tilde{\mathcal{E}}(\tilde{t})$ *be the Nash equilibrium set for* $\tilde{\Gamma}(\tilde{t})$.

1. *For every* $\hat{t} \prec \tilde{t}$, *if* $x^* \preceq \underline{B}(\tilde{y}, \tilde{t})$, *then* $\tilde{\mathcal{E}}(\tilde{t}) \subset \mathcal{E}(\tilde{t})$ *and for every* $y^* \in \tilde{\mathcal{E}}(\tilde{t})$, $x^* \preceq y^*$.
2. *For every* $\hat{t} \prec \tilde{t}$, *if* $x^* \prec \underline{B}(\tilde{y}, \tilde{t})$, *then* $\tilde{\mathcal{E}}(\tilde{t}) \subset \mathcal{E}(\tilde{t})$ *and for every* $y^* \in \tilde{\mathcal{E}}(\tilde{t})$, $x^* \prec y^*$.
3. *For every* $\hat{t} \prec \tilde{t}$, *if for every* i, $x_i^* \prec \underline{B}_i(\tilde{y}_{-i}, \tilde{t})$, *then* $\tilde{\mathcal{E}}(\tilde{t}) \subset \mathcal{E}(\tilde{t})$ *and for every* $y^* \in \tilde{\mathcal{E}}(\tilde{t})$ *and for every* i, $x_i^* \prec y_i^*$.

Proof Consider $\hat{t} \prec \tilde{t}$, $x^* \in \mathcal{E}(\hat{t})$, and let $\tilde{y} = \overline{B}(x^*, \tilde{t})$. B is (weakly) increasing in t implies $B(x^*, \hat{t}) \sqsubseteq B(x^*, \tilde{t})$, and therefore, $x^* \preceq \overline{B}(x^*, \hat{t}) \preceq \overline{B}(x^*, \tilde{t}) = \tilde{y}$. Consider the nonempty interval $[x^*, \tilde{y}]$, let $\tilde{\Gamma}(\tilde{t})$ be the lattice game at $\tilde{t}$ restricted to $[x^*, \tilde{y}]$, and let $\tilde{\mathcal{E}}(\tilde{t})$ be the Nash equilibrium set for $\tilde{\Gamma}(\tilde{t})$. If $\tilde{\mathcal{E}}(\tilde{t}) = \emptyset$, each of the three statements is vacuously true.

The hypothesis in statement (1) implies that for every $z \in [x^*, \tilde{y}]$, $B(z, \tilde{t}) \subset [x^*, \tilde{y}]$, and this is seen as follows. First, $z \preceq \tilde{y}$ implies $B(\tilde{y}, \tilde{t}) \sqsubseteq B(z, \tilde{t})$, and therefore, $x^* \preceq \underline{B}(\tilde{y}, \tilde{t}) \preceq \underline{B}(z, \tilde{t})$, where the first inequality follows from the hypothesis in condition (1). Second, $x^* \preceq z$ implies $B(z, \tilde{t}) \sqsubseteq B(x^*, \tilde{t})$, and therefore, $\overline{B}(z, \tilde{t}) \preceq \overline{B}(x^*, \tilde{t}) = \tilde{y}$. Taken together, it follows that $B(z, \tilde{t}) \subset [x^*, \tilde{y}]$. In other words, for every $z \in [x^*, \tilde{y}]$ and for every i, $B_i(z_{-i}, \tilde{t}) \subset [x_i^*, \tilde{y}_i]$, and therefore, $B(z, \tilde{t}) \subset [x^*, \tilde{y}]$.

Consider the restricted game $\tilde{\Gamma}(\tilde{t})$. For every $z \in [x^*, \tilde{y}]$ and for every player i, let $\tilde{B}_i(z_{-i}, \tilde{t})$ denote the best response correspondence of player i in the restricted game, that is, $\tilde{B}_i(z_{-i}, \tilde{t}) = \arg\max_{\xi \in [x_i^*, \tilde{y}_i]} u_i(\xi, z_{-i}, \tilde{t})$. Notice that for every $z \in [x^*, \tilde{y}]$ and for every player i, $\tilde{B}_i(z_{-i}, \tilde{t}) = B_i(z_{-i}, \tilde{t})$, as follows. If $x_i \in B_i(z_{-i}, \tilde{t})$, then for every $\xi \in X_i$, $u_i(x_i, z_{-i}, \tilde{t}) \geq u_i(\xi_i, z_{-i}, \tilde{t})$, and therefore, for every $\xi \in [x_i^*, \tilde{t}_i]$, $u_i(x_i, z_{-i}, \tilde{t}) \geq u_i(\xi_i, z_{-i}, \tilde{t})$, implying $\tilde{B}_i(z_{-i}, \tilde{t}) \subset B_i(z_{-i}, \tilde{t})$. In the other direction, if $x_i \in \tilde{B}_i(z_{-i}, \tilde{t})$, then for every $\xi \in [x_i^*, \tilde{t}_i]$, $u_i(x_i, z_{-i}, \tilde{t}) \geq u_i(\xi_i, z_{-i}, \tilde{t})$, and as $B_i(x_{-i}, \tilde{t}) \subset [x_i^*, \tilde{y}_i]$, the maximum is achieved in $[x_i^*, \tilde{y}_i]$, and therefore, for every $\xi \in X_i$, $u_i(x_i, z_{-i}, \tilde{t}) \geq u_i(\xi_i, z_{-i}, \tilde{t})$, implying $\tilde{B}_i(z_{-i}, \tilde{t}) \subset B_i(z_{-i}, \tilde{t})$.

Let $\tilde{B}(z, \tilde{t})$ denote the joint best response correspondence in the restricted game. The previous argument shows that for every $z \in [x^*, \tilde{y}]$, $\tilde{B}(z, \tilde{t}) = B(z, \tilde{t})$, and therefore, every fixed point of $\tilde{B}(\cdot, \tilde{t})$ is a fixed point of $B(\cdot, \tilde{t})$, and this implies $\tilde{\mathcal{E}}(\tilde{t}) \subset \mathcal{E}(\tilde{t})$. Moreover, for every $y^* \in \tilde{\mathcal{E}}(\tilde{t}) \subset \mathcal{E}(\tilde{t})$, $y^* \preceq \tilde{y}$

and therefore, the hypothesis in statement (1) and strategic substitutes imply $x^* \preceq \underline{B}(\tilde{y}, \tilde{t}) \preceq \underline{B}(y^*, \tilde{t}) \preceq y^*$.

Statement (2) follows from the same argument and modifying the last sentence in the argument with the hypothesis in statement (2), that is, $x^* \prec \underline{B}(\tilde{y}, \tilde{t}) \preceq \underline{B}(y^*, \tilde{t}) \preceq y^*$. Statement (3) follows from the same argument as well, modifying the last sentence in the argument with the hypothesis in statement (3), that is, for every i, $x_i^* \prec \underline{B}_i(\tilde{y}_{-i}, \tilde{t}) \preceq \underline{B}_i(y_{-i}^*, \tilde{t}) \preceq y_i^*$. □

This theorem is formulated without invoking applicability of a particular fixed point theorem or other device to guarantee existence of equilibrium. In a parameterized lattice game, when Nash equilibrium set is nonempty on intervals, then $\tilde{\mathcal{E}}(\tilde{t}) \neq \emptyset$, and monotone comparative statics of equilibrium outcomes follows.

Theorem 4.40 (MCS-Parameterized GSS, part 2) *In a parameterized GSS, for each $\hat{t}, \tilde{t} \in T$ with $\hat{t} \prec \tilde{t}$, and for each $x^* \in \mathcal{E}(\hat{t})$, let $\tilde{y} = \overline{B}(x^*, \tilde{t})$, let $\tilde{\Gamma}(\tilde{t})$ be the lattice game at $\tilde{t}$ restricted to $[x^*, \tilde{y}]$, let $\tilde{\mathcal{E}}(\tilde{t})$ be the Nash equilibrium set for $\tilde{\Gamma}(\tilde{t})$, and suppose $\tilde{\mathcal{E}}(\tilde{t}) \neq \emptyset$.*

1. *If $x^* \preceq \underline{B}(\tilde{y}, \tilde{t})$, then there is $y^* \in \mathcal{E}(\tilde{t})$ such that $x^* \preceq y^*$.*
2. *If $x^* \prec \underline{B}(\tilde{y}, \tilde{t})$, then there is $y^* \in \mathcal{E}(\tilde{t})$ such that $x^* \prec y^*$.*
3. *If for every i, $x_i^* \prec \underline{B}_i(\tilde{y}_{-i}, \tilde{t})$, then there is $y^* \in \mathcal{E}(\tilde{t})$ such that for every i, $x_i^* \prec y_i^*$.*

Proof As $\tilde{\mathcal{E}}(\tilde{t}) \neq \emptyset$, let $y^* \in \tilde{\mathcal{E}}(\tilde{t})$. By the previous theorem, $y^* \in \mathcal{E}(\tilde{t})$ and the result follows by applying the corresponding statement in Theorem 4.39 (page 124) to this y^*. □

Theorem 4.40 (page 125) generalizes the result due to Roy and Sabarwal (2010) for the special case of convex action spaces and convex valued best response correspondence.

The condition $x^* \preceq \underline{B}(\tilde{y}, \tilde{t})$ may be seen graphically when there are two players, each with singleton valued best response. In that case, $x^* \preceq \underline{B}(\tilde{y}, \tilde{t})$ is equivalent to $x_1^* \preceq B_1(B_2(x_1^*, \tilde{t}), \tilde{t})$ and $x_2^* \preceq B_2(B_1(x_2^*, \tilde{t}), \tilde{t})$.

Figure 4.2 (page 126) shows the case when both conditions hold. For both players, the direct parameter effect providing an incentive to take a higher action is sufficient to overcome the indirect strategic substitute effect driving a lower action. In the new equilibrium, both players play a higher action.

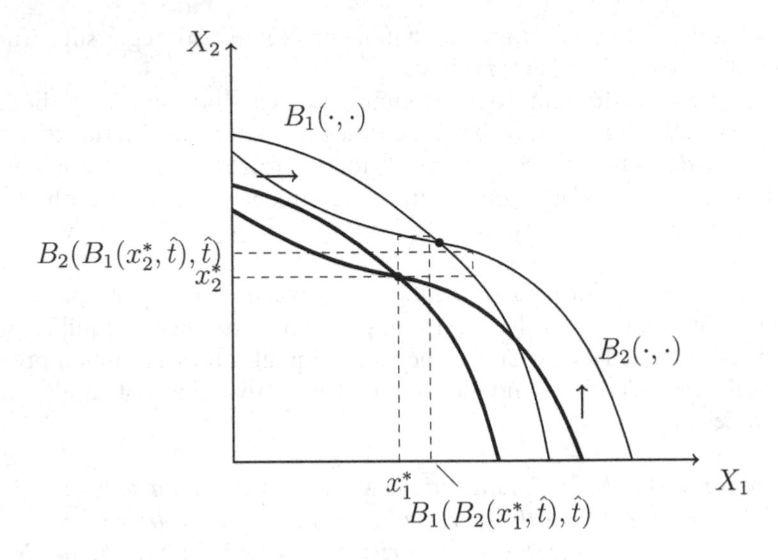

Fig. 4.2 Favorable composition of direct and indirect effects

Figure 4.3 (page 127) shows the case when only one condition holds, say, $x_1^* \npreceq B_1(B_2(x_1^*, \tilde{t}), \tilde{t})$, but $x_2^* \preceq B_2(B_1(x_2^*, \tilde{t}), \tilde{t})$. For player 1, the direct parameter effect providing an incentive to take a higher action is insufficient to counteract the indirect strategic substitute effect driving a lower action. In the new equilibrium, player 1 action goes down.

More generally, when best response of each player is a function, the condition $x^* \preceq \underline{B}(\tilde{y}, \tilde{t})$ is equivalent to $B(B(x^*, \hat{t}), \hat{t}) = x^* \preceq B(B(x^*, \tilde{t}), \tilde{t})$, and this is equivalent to $B(B(x^*, t), t)$ is (weakly) increasing in t at $\hat{t}$.

When payoff functions are twice continuously differentiable, the condition $B(B(x^*, t), t)$ is (weakly) increasing in t at $\hat{t}$ may be formulated in terms of an equivalent condition on payoff functions, making it more accessible and easier to apply, as shown in Roy and Sabarwal (2010). In the following theorem, for notational convenience, superscript is used to index payoff function of player i, denoted u^i, and subscripts to denote partial derivatives. That is, $u^i_{i,t} = \frac{\partial^2 u^i}{\partial t \partial x_i}(x_1, \ldots, x_I, t)$, $u^i_{i,j} = \frac{\partial^2 u^i}{\partial x_j \partial x_i}(x_1, \ldots, x_I, t)$, and so on.

Theorem 4.41 (MCS-Differentiable payoffs) *Consider a parameterized GSS in which the parameter space is an interval in the reals, and for every*

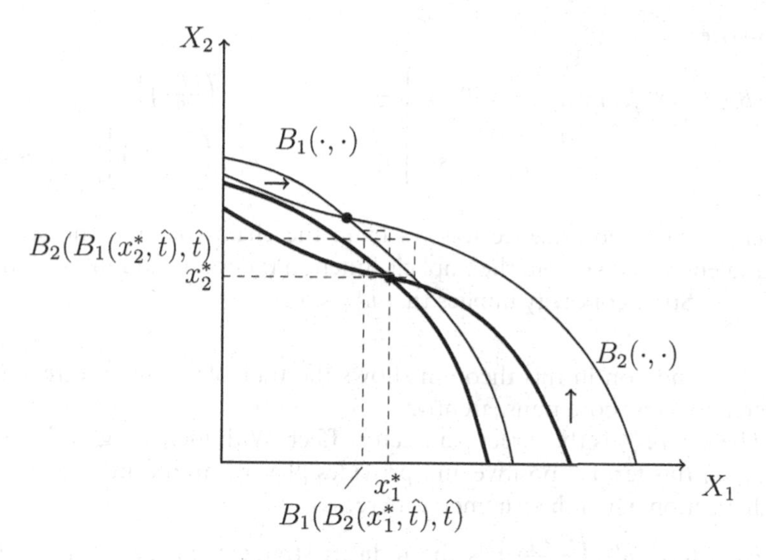

Fig. 4.3 Unfavorable composition of direct and indirect effects

player i, their action space is an interval in the reals, their payoff is twice continuously differentiable, and their payoff is strictly concave in own variable. For $\hat{t}$ in interior of T and $x^* \in \mathcal{E}(\hat{t})$, the following are equivalent.

1. *The composition $B(B(x^*, t), t)$ is increasing in t at $\hat{t}$.*
2. *For every player i,*

$$\left[u^i_{i,t} + \sum_{j \neq i} u^i_{i,j} \left(-\frac{u^j_{j,t}}{u^j_{j,j}} \right) \right] \Bigg|_{(x^*,\hat{t})} > 0.$$

Proof Payoff of each player is strictly concave in own variable implies that best response of each player is singleton valued. Statement (1) is equivalent to the statement that for every player i, $\frac{\partial}{\partial t} B_i(B_{-i}(x^*, t), t)\big|_{(x^*,t)} > 0$, where $B_{-i}(x^*, t)$ is the joint best response of players other than i to x^* at t.

Moreover,

$$\frac{\partial}{\partial t} B_i(B_{-i}(x^*_{-i}, t), t)\Big|_{(x^*,t)} > 0 \Leftrightarrow \left[\frac{\partial B_i}{\partial t} + \sum_{j \neq i} \frac{\partial B_i}{\partial x_j}\left(\frac{\partial B_j}{\partial t}\right)\right]\Big|_{(x^*,t)} > 0$$
$$\Leftrightarrow \left[u^i_{i,t} + \sum_{j \neq i} u^i_{i,j}\left(-\frac{u^j_{j,t}}{u^j_{j,j}}\right)\right]\Big|_{(x^*,\hat{t})} > 0,$$

where the first equivalence follows from the chain rule and the second equivalence follows from the implicit function theorem and multiplication by $-\frac{1}{u^i_{i,i}}$. Strict concavity implies that $u^i_{i,i} < 0$. □

The condition in this theorem shows the tradeoff between direct and indirect effects more transparently.

The term $u^i_{i,t}$ is the direct parameter effect. With increasing differences in (x_i, t) this term is positive and provides player i an incentive to take a higher action when the parameter increases.

Each term $u^i_{i,j}\left(-\frac{u^j_{j,t}}{u^j_{j,j}}\right)$ is the indirect strategic substitute effect that opponent j has on payoff of player i. With increasing differences in (x_j, t), decreasing differences in (x_i, x_j), and payoff that is strictly concave in own variable, each such term is negative and drives player i to take a lower action when player j action goes up due to an increase in the parameter. The total indirect strategic substitutes effect is the sum of these effects for opponents $j \neq i$.

The idea that composition of these two opposing effects is favorable is formalized by the condition that the sum of the direct and indirect effects is positive.

The theorem generalizes immediately to the case in which the parameter space is a poset with nonempty interior in finite-dimensional Euclidean space, and for every player i, their action space is a subcomplete lattice with nonempty interior in finite-dimensional Euclidean space, their payoff is twice continuously differentiable, and their payoff is strictly concave in own variable.

Example 4.42 (Parameterized asymmetric Cournot duopoly, continued) In asymmetric Cournot duopoly (Example 4.38, page 119), payoffs are given by

$$u^1(x_1, x_2) = (a - b(x_1 + x_2))x_1 - (c - \alpha t)x_1$$
$$u^2(x_1, x_2) = (a - b(x_1 + x_2))x_2 - (c - (1 - \alpha)t)x_2.$$

For firm 1, $u^1_{1,t} = \alpha$, $u^1_{1,2} = -b$, $u^2_{2,t} = 1 - \alpha$, and $u^2_{2,2} = -2b$, and the condition in Theorem 4.41 (page 126) is $\alpha + (-b)\left(-\frac{1-\alpha}{-2b}\right) > 0$, which holds exactly when $\alpha > \frac{1}{3}$. For firm 2, $u^2_{2,t} = 1 - \alpha$, $u^2_{2,1} = -b$, $u^1_{1,t} = \alpha$, and $u^1_{1,1} = -2b$, and the condition in Theorem 4.41 (page 126) is $(1 - \alpha) + (-b)\left(-\frac{\alpha}{-2b}\right) > 0$, which holds exactly when $\alpha < \frac{2}{3}$. These conditions do not depend on the equilibrium, and therefore, the equilibrium profile increases monotonically when $\frac{1}{3} < \alpha < \frac{2}{3}$.

Example 4.43 (Parameterized worker competition) Consider the worker competition game (Example 4.34, page 112) and consider environments in which expected benchmark effort in the firm may be different. For example, perhaps benchmark effort is higher in some industries such as stock market trading or start-up firms. Or, perhaps a firm is considering raising standards by introducing higher benchmarks and may provide some implicit benefits to employees, for example, free cafeteria food or free tuition. Suppose these environments are summarized by a reduced form benchmark parameter $t \in [\frac{1}{2}, 1]$, where $t = \frac{1}{2}$ was the case earlier. Payoff for worker i is given by

$$u_i(x_1, x_2, x_3, t) = 1 - x_1x_2x_3 - (x_i - t)^2.$$

The condition in Theorem 4.41 (page 126) is computed as follows. For worker i, $u^i_{i,t} = 2$, $u^i_{i,j} = -x_k$, $u^i_{i,k} = -x_j$, $u^j_{j,t} = u^k_{k,t} = 2$, and $u^j_{j,j} = u^k_{k,k} = -2$. Therefore, for worker i, the condition is $2 - x_j - x_k$ and this is positive as long as one of the other workers is not exerting maximum possible intensity of effort. In particular, this condition is satisfied at $t = \frac{1}{2}$ with equilibrium (0.42, 0.42, 0.42) showing that an increase in benchmark effort causes each worker to put in greater effort in the new equilibrium.

For completeness, parameterized best response of worker i is given by $B_i(x_j, x_k, t) = t - \frac{1}{2}x_jx_k$. This is increasing in t, (weakly) decreasing in (x_j, x_k) and strictly decreasing when $x_j, x_k > 0$. The unique Nash equilibrium is given by $(\sqrt{1+2t} - 1, \sqrt{1+2t} - 1, \sqrt{1+2t} - 1)$.

This chapter develops the theory of games with strategic substitutes. Results for this class of games may be very different from those for games with strategic complements. These results show new insights for interdependent and equilibrium behavior of collections of players with strategic substitutes. Several results generalize to the larger class of lattice games.

References

Acemoglu, D., & Jensen, M. K. (2013). Aggregate Comparative Statics. *Games and Economic Behavior, 81*, 27–49.

Amir, R. (1996). Cournot Oligopoly and the Theory of Supermodular Games. *Games and Economic Behavior, 15*, 132–148.

Bulow, J. I., Geanakoplos, J. D., & Klemperer, P. D. (1985). Multimarket Oligopoly: Strategic Substitutes and Complements. *Journal of Political Economy, 93*(3), 488–511.

Cao, Z., Chen, X., Qin, C.-Z., Wang, C., & Yang, X. (2018). Embedding Games with Strategic Complements into Games with Strategic Substitutes. *Journal of Mathematical Economics, 78*, 45–51.

Dubey, P., Haimanko, O., & Zapechelnyuk, A. (2006). Strategic Complements and Substitutes, and Potential Games. *Games and Economic Behavior, 54*, 77–94.

Fudenberg, D., & Tirole, J. (1984). The Fat-Cat Effect, the Puppy-Dog Ploy, and the Lean and Hungry Look. *American Economic Review, 74*(2), 361–366.

Jensen, M. K. (2010). Aggregative Games and Best-Reply Potentials. *Economic Theory, 43*(1), 45–66.

Monaco, A., & Sabarwal, T. (2016). Games with Strategic Complements and Substitutes. *Economic Theory, 62*(1), 65–91.

Monderer, D., & Shapley, L. (1996). Potential Games. *Games and Economic Behavior, 14*, 124–143.

Rosenthal, R. W. (1973). A Class of Games Possessing Pure-Strategy Nash Equilibria. *International Journal of Game Theory, 2*(1), 65–67.

Roy, S., & Sabarwal, T. (2008). On the (Non-)Lattice Structure of the Equilibrium Set in Games With Strategic Substitutes. *Economic Theory, 37*(1), 161–169.

Roy, S., & Sabarwal, T. (2010). Monotone Comparative Statics for Games with Strategic Substitutes. *Journal of Mathematical Economics, 46*(5), 793–806.

Roy, S., & Sabarwal, T. (2012). Characterizing Stability Properties in Games with Strategic Substitutes. *Games and Economic Behavior, 75*(1), 337–353.

Zimper, A. (2007). A Fixed Point Characterization of the Dominance-Solvability of Lattice Games with Strategic Substitutes. *International Journal of Game Theory, 36*(1), 107–117.

CHAPTER 5

Monotone Games

Abstract Monotone games is an umbrella that encompasses games with strategic complements, games with strategic substitutes, and combinations of the two. The scope of each of these categories is large and each has similarities to others and differences as well. Previous chapters are devoted to monotone games with strategic complements and monotone games with strategic substitutes. This chapter develops the general model that encompasses these two and focuses on study of games in which both types of players are present simultaneously.

Keywords Monotone incentives · Strategic complements and substitutes · Totally unordered equilibrium set · Never decreasing equilibrium

A broader impact of the newer and direct study of GSS is that it has opened the path to examine more general cases.

Earlier chapters study games in which either all players have strategic complements (GSC) or all players have strategic substitutes (GSS). Each of these classes of games is a subclass of lattice games. This chapter develops the theory of games in which both types of players are present simultaneously. This is a subclass of lattice games as well.

As shown in Chapter 4 (page 79), several results for GSS generalize to lattice games. Therefore, as a special case, those results apply as stated to

T. Sabarwal, *Monotone Games*,
https://doi.org/10.1007/978-3-030-45513-2_5

the class of lattice games in which some players have strategic complements and others have strategic substitutes. In such games, the Nash equilibrium set may be empty, and under mild conditions on one or two players only, the Nash equilibrium set is totally unordered. This implies the same result for uniqueness of symmetric Nash equilibrium on action spaces that are chains.

Moreover, distinctive features of games with both types of players imply new patterns for individual and equilibrium behavior. When both types of players are present, the behavior of the joint best response dynamic is more complex than in the case in which all players are of one type only. This is accounted for in an appropriately adjusted notion of lower response dynamics and upper response dynamics. Limits of these dynamics provide bounds on serially undominated strategies and adaptive dynamics and yield a corresponding result on equivalence of dominance solvability and global stability. Uniqueness of Nash equilibrium is not a sufficient condition for this result, making it similar to the case for GSS.

Monotone comparative statics of equilibrium outcomes is easier to show in the two player case and harder more generally. When there are two players (one player with strategic complements and one with strategic substitutes), action spaces are chains, best responses are functions, and there are parametric incentives to take higher actions, the player with strategic complements always takes a higher equilibrium action at a higher parameter value. The player with strategic substitutes takes a higher equilibrium action, if, and only if, the tradeoff between direct and indirect effects is satisfied in the same manner as for the case of strategic substitutes in Chapter 4.

With more than two players, there are sufficient conditions. These conditions may be motivated in terms of direct and indirect effects similar to the case for strategic substitutes in Chapter 4. These conditions are needed only for players with strategic substitutes. No additional conditions are needed for players with strategic complements. This presents an appropriately intuitive blend of results for both parameterized GSC and parameterized GSS.

5.1 Monotone Incentives

Strategic complements formalize environments in which decision-making is governed by incentives to move in the same direction as others. This is the case of codirectional incentives.

Strategic substitutes formalize environments in which decision-making is governed by incentives to move in the direction opposite to others. This is the case of contradirectional incentives.

There are situations in which both types of incentives are present. For example, in law enforcement, police want to be at the same place as a criminal (coordination), but criminal wants to be in a place different from police (opposition). The same dynamic is present between advertisers and consumers, a baseball pitcher and a hitter, hunter and hunted, dictator and rebel, and so on. These situations exhibit patterns of both coordination and opposition. In these situations, some players have strategic complements and others have strategic substitutes.

Collectively, these three are categories with ***monotone interdependent incentives***, or more briefly, ***monotone incentives***.

More generally, the focus here is on situations in which a subset of the decision-makers have codirectional incentives for coordination and the remaining decision-makers have contradirectional incentives for opposing moves. When one or the other subset is empty, the results specialize to those that hold in the corresponding situation analyzed earlier. When neither subset is empty, the additional results developed here apply.

When monotone incentives take the form of codirectional incentives, results about individual behavior and equilibrium may be imported from Chapter 3 (page 45), and when monotone incentives take the form of contradirectional incentives, results about individual behavior and equilibrium may be imported from Chapter 4 (page 79). The interdependence of actions of one type on actions of the other type gives rise to new patterns of behavior studied in more detail here.

5.2 Monotone Game

5.2.1 *Definition*

A ***monotone game*** is a lattice game in which for every player i, either player i has strategic complements or player i has strategic substitutes. In other words, a monotone game is a lattice game $\Gamma = (X_i, u_i)_{i=1}^{I}$, where

1. There are finitely many players, indexed by $i \in \{1, \ldots, I\}$
2. For each i, the action space of player i is X_i, where X_i is a nonempty, subcomplete lattice in $\mathbb{R}^{n_i}$ with the Euclidean order.
3. For each i, payoff for player i is $u_i : X_i \times X_{-i} \to \mathbb{R}$, where

(a) u_i is upper semicontinuous on X and continuous on X_{-i}, and
(b) For each x_{-i}, $u_i(\cdot, x_{-i})$ is quasisupermodular on X_i, and either u_i has single crossing property in (x_i, x_{-i}) or u_i has dual single crossing property in (x_i, x_{-i}).

Monotone games with both strategic complements and strategic substitutes are discussed in the early work of Fudenberg and Tirole (1984) and Bulow et al. (1985). Results for the general case of lattice games and monotone games are available in Roy and Sabarwal (2008), Roy and Sabarwal (2010), Roy and Sabarwal (2012), Monaco and Sabarwal (2016), and Barthel and Hoffmann (2019). Aggregative models include Rosenthal (1973), Monderer and Shapley (1996), Dubey et al. (2006), Jensen (2010), and Acemoglu and Jensen (2013).

As earlier, although a player's action space is assumed to be finite dimensional, the material on codirectional incentives and contradirectional incentives in earlier chapters is developed in its natural and more general setting. Moreover, proofs of results in this section are developed in a manner that shows their natural extension to more general settings with additional assumptions.

Player i is a ***strategic complements player***, if u_i has single crossing property in (x_i, x_{-i}), and is a ***strategic substitutes player***, if u_i has dual single crossing property in (x_i, x_{-i}). In the degenerate case when u_i has both single crossing property in (x_i, x_{-i}) and dual single crossing property in (x_i, x_{-i}), we term the player a strategic complements player.

Let C denote the collection of strategic complements players and S the collection of strategic substitutes players. This partitions the set of all players into disjoint subsets C and S, with $C \cup S = \{1, \ldots, I\}$. When $S = \emptyset$, a monotone game specializes to a game with strategic complements, and when $C = \emptyset$, it specializes to a game with strategic substitutes.

Let $X_C = \times_{i \in C} X_i$ denote the space of profiles of actions of strategic complements players and $X_S = \times_{i \in S} X_i$ denote the space of profiles of actions of strategic substitutes players. When convenient, we denote $X = X_C \times X_S$.

The definition implies immediately that every GSC is a monotone game ($S = \emptyset$) and every GSS is a monotone game ($C = \emptyset$). All the games in the introduction (Chapter 1) are monotone games. Matching pennies (Example 1.3, page 4) is a monotone game that is neither a game with strategic complements nor a game with strategic substitutes.

Table 5.1 Online content provision

		P2			
		L	ML	MH	H
P1	L	1, 4	1, 4	4, 2	4, 1
	ML	2, 1	4, 4	4, 4	3, 1
	MH	4, 1	4, 2	3, 4	2, 4
	H	4, 1	3, 2	1, 3	1, 4

Example 5.1 (Online content provision) Consider an interaction between online content providers (such as newspapers, magazines, podcast developers, online news organizations, multimedia providers, and so on) and consumers.

An online content provider chooses quality of content to provide. Lower quality content relies more on advertising, higher quality content may be monetized in other ways. Success of advertising depends on use of ad blocking software. When ad blocking is lower, content provides are better off with lower quality content and more advertising. When ad blocking is higher it is better to provide higher quality content that can be monetized in other ways.

An online consumer controls intensity of advertisement delivery by using ad blocking software. If content quality is lower, it is better for consumer to use higher ad blocking than when content quality is higher.

Suppose there are four content quality levels and four ad blocking settings, each given by low (L), medium-low (ML), medium-high (MH), and high (H), with $L \prec ML \prec MH \prec H$. Payoffs are given in Table 5.1 (page 135), with the consumer denoted as player 1 and the content provider denoted as player 2. It can be checked that consumer payoff satisfies dual single crossing property and content provider payoff satisfies single crossing property. Therefore, the consumer has strategic substitutes and the content provider has strategic complements.

Example 5.2 (Bernot duopoly) Consider an industry with two firms producing branded handbags. One firm focuses more on selling its product by choosing price and the other focuses more on selling its product by choosing quantity of product to sell. In other words, one firm competes as in the Bertrand case and the other firm competes as in the Cournot case. This is termed a Bernot duopoly here.

Suppose inverse market demand for firm 1 is given by $p_1 = a - 2q_1 - q_2$ and for firm 2 by $p_2 = a - 2q_2 - q_1$, where p_1 is price of firm 1 handbags, p_2 is price of firm 2 handbags, q_1 is output of firm 1 handbags, q_2 is output of firm 2 handbags, and $a > 0$ is an index for strength of demand. Handbags are produced at constant marginal cost $c > 0$. Firm 1 chooses p_1 and firm 2 chooses q_2. Profit of each firm is given by total revenue minus total cost.

Using the decision variables (p_1, q_2), profit of firm 1 is given by $\pi_1(p_1, q_2) = (p_1 - c)\frac{1}{2}(a - q_2 - p_1)$. This function satisfies dual single crossing property in (p_1, q_2), because $\frac{\partial^2 \pi_1}{\partial q_2 \partial p_1} = -\frac{1}{2} < 0$. Therefore, firm 1 has strategic substitutes.

Profit of firm 2 is given by $\pi_2(p_1, q_2) = (a - \frac{1}{2}(a - q_2 - p_1) - 2q_2 - c)q_2$. This function satisfies single crossing property in (p_1, q_2), because $\frac{\partial^2 \pi_2}{\partial q_2 \partial p_1} = \frac{1}{2} > 0$, and therefore, firm 2 has strategic complements.

5.2.2 *Monotone Best Response*

Recall that in a lattice game $\Gamma = (X_i, u_i)_{i=1}^I$, for each player i, best response set of player i to opponent actions x_{-i} is $B_i(x_{-i}) = \arg\max_{x_i \in X_i} u_i(x_i, x_{-i})$, the best response correspondence for player i is $B_i : X_{-i} \rightrightarrows X_i$, mapping x_{-i} to $B_i(x_{-i})$, and the joint best response correspondence is $B : X \rightrightarrows X$, given by $B(x) = B_1(x_{-1}) \times B_2(x_{-2}) \times \cdots \times B_I(x_{-I})$.

For a given profile x, the ***joint best response of strategic complements players*** is $B_C(x) = \times_{i \in C} B_i(x_{-i})$, and the ***joint best response correspondence of strategic complements players*** is $B_C : X \rightrightarrows X_C$, mapping x to $B_C(x)$.

Similarly, the ***joint best response of strategic substitutes players*** is $B_S(x) = \times_{i \in S} B_i(x_{-i})$, and the ***joint best response correspondence of strategic substitutes players*** is $B_S : X \rightrightarrows X_S$, mapping x to $B_S(x)$. When convenient, we denote the joint best response of all players as $B(x) = B_C(x) \times B_S(x)$.

In the more general framework of a lattice game the best response correspondence of each player and the joint best response correspondence are nonempty valued, compact valued, and upper hemicontinuous (Theorem 2.26, page 32). Moreover, Theorem 3.17 (page 63) shows that for strategic complements players, they are subcomplete valued and increasing, and Theorem 4.16 (page 95) shows that for strategic substitutes players, they are subcomplete valued and decreasing.

These properties at the level of an individual player can be combined naturally to understand joint behavior of best responses when some players have strategic complements and others have strategic substitutes providing a basis to analyze behavior in the more complex monotone game with both types of players.

Theorem 5.3 *In every monotone game* $\Gamma = (X_i, u_i)_{i=1}^I$,

1. *The joint best response correspondence of strategic complements players is nonempty valued, compact valued, subcomplete valued, upper hemicontinuous, and (weakly) increasing.*
2. *The joint best response correspondence of strategic substitutes players is nonempty valued, compact valued, subcomplete valued, upper hemicontinuous, and (weakly) decreasing.*

Proof Statement (1) follows from Theorem 3.17 (page 63) and statement (2) from Theorem 4.16 (page 95). □

Examples 5.4 Consider the following examples.

1. In matching pennies (Example 1.3, page 4), best response of player 1 is decreasing, best response of player 2 is increasing, and both best responses are singleton valued.
2. In online content provision (Example 5.1, page 135), best responses are as follows.

$$\begin{aligned} B_1(L) &= \{MH, H\}, & B_1(ML) &= \{ML, MH\}, \\ B_1(MH) &= \{L, ML\}, & B_1(H) &= \{L\}, \\ B_2(L) &= \{L, ML\}, & B_2(ML) &= \{ML, MH\}, \\ B_2(MH) &= \{MH, H\}, & B_2(H) &= \{H\}. \end{aligned}$$

   Best response of player 1 is (weakly) decreasing, best response of player 2 is (weakly) increasing, and both best responses have non-singleton values.
3. In Bernot duopoly (Example 5.2, page 135), best response of firm 1 is given by $B_1(q_2) = \frac{1}{2}(a+c-q_2)$ and best response of firm 2 is given by $B_2(p_1) = \frac{1}{6}(a - 2c + p_1)$. Best response of firm 1 is decreasing in opponent choice, best response of firm 2 is increasing in opponent choice, and both best responses are functions.

When some players have (weakly) decreasing best response correspondence and others have (weakly) increasing best response correspondence, the behavior patterns become more complex than in either GSS or GSC. In particular, neither the version of best response dynamics in GSC nor the mixtures version in GSS are helpful in this case. A notion that simultaneously combines what was used in GSC to account for strategic complements players and what was used in GSS to account for strategic substitutes players is needed.

In a monotone game, the ***lower response dynamic***, $(\underline{x}^n)_{n=0}^{\infty}$, and the ***upper response dynamic***, $(\overline{x}^n)_{n=0}^{\infty}$, are defined as follows. Let $\underline{x}^0 = \inf X$ and $\overline{x}^0 = \sup X$, and for $n \geq 1$, let

$$\underline{x}^n = (\underline{B}_C(\underline{x}^{n-1}), \underline{B}_S(\overline{x}^{n-1})) \quad \text{and} \quad \overline{x}^n = (\overline{B}_C(\overline{x}^{n-1}), \overline{B}_S(\underline{x}^{n-1})).$$

Here, $\underline{B}_C(\underline{x}^{n-1}) = \inf B_C(\underline{x}^{n-1})$, $\underline{B}_S(\overline{x}^{n-1}) = \inf B_S(\overline{x}^{n-1})$, $\overline{B}_C(\overline{x}^{n-1}) = \sup B_C(\overline{x}^{n-1})$, and $\overline{B}_S(\underline{x}^{n-1}) = \sup B_S(\underline{x}^{n-1})$.

When there are no strategic substitutes players ($S = \emptyset$), the monotone game is a GSC, and this definition specializes, respectively, to that of best response dynamic from inf X and from sup X in a GSC. When there are no strategic complements players ($C = \emptyset$), the monotone game is a GSS, and this definition specializes, respectively, to that of lower mixture and upper mixture of best response dynamics in a GSS.

This definition applies more generally when there are both types of players. It accounts for the different incentives for coordination and opposition arising from presence of both types of players. The lower and upper response dynamics are useful in this case. This may be seen in the next result, due to Barthel and Hoffmann (2019). The continuity properties on payoffs here are weaker than the ones they used and upper hemicontinuity follows from the more general result for lattice games.

Theorem 5.5 *In a monotone game, let* $(\underline{x}^n)_{n=0}^{\infty}$ *and* $(\overline{x}^n)_{n=0}^{\infty}$ *be the lower and upper response dynamics, respectively.*

1. *For every* n, $\underline{x}^n \preceq \underline{x}^{n+1} \preceq \overline{x}^{n+1} \preceq \overline{x}^n$.
2. *Let* $\underline{x} = \bigvee_{n=0}^{\infty} \underline{x}^n$ *and* $\overline{x} = \bigwedge_{n=0}^{\infty} \overline{x}^n$. *Then* $\underline{x} \preceq \overline{x}$.
3. *For every* $i \in C$, $\underline{x}_i \in B_i(\underline{x}_{-i})$ *and* $\overline{x}_i \in B_i(\overline{x}_{-i})$.
4. *For every* $i \in S$, $\underline{x}_i \in B_i(\overline{x}_{-i})$ *and* $\overline{x}_i \in B_i(\underline{x}_{-i})$.

Proof For statement (1), we use induction. For $n = 0$, $\underline{x}^0 = \inf X \preceq \sup X = \overline{x}^0$. Therefore, for every $i \in C$, $\underline{x}_i^0 \preceq \underline{B}_i(\underline{x}_{-i}^0) \preceq \overline{B}_i(\overline{x}_{-i}^0) \preceq \overline{x}_i^0$, and for every $i \in S$, $\underline{x}_i^0 \preceq \underline{B}_i(\overline{x}_{-i}^0) \preceq \overline{B}_i(\underline{x}_{-i}^0) \preceq \overline{x}_i^0$. Consequently, $\underline{x}^0 \preceq \underline{x}^1 \preceq \overline{x}^1 \preceq \overline{x}^0$. Now, suppose $\underline{x}^n \preceq \underline{x}^{n+1} \preceq \overline{x}^{n+1} \preceq \overline{x}^n$. Then for every $i \in C$,

$$\underline{x}_i^{n+1} = \underline{B}_i(\underline{x}_{-i}^n) \preceq \underline{B}_i(\underline{x}_{-i}^{n+1}) \preceq \overline{B}_i(\overline{x}_{-i}^{n+1}) \preceq \overline{B}_i(\overline{x}_{-i}^n) = \overline{x}_i^{n+1},$$

and for every $i \in S$,

$$\underline{x}_i^{n+1} = \underline{B}_i(\overline{x}_{-i}^n) \preceq \underline{B}_i(\overline{x}_{-i}^{n+1}) \preceq \overline{B}_i(\underline{x}_{-i}^{n+1}) \preceq \overline{B}_i(\underline{x}_{-i}^n) = \overline{x}_i^{n+1}.$$

Consequently, $\underline{x}^n \preceq \underline{x}^{n+1} \preceq \overline{x}^{n+1} \preceq \overline{x}^n$, proving statement (1). Statement (2) follows immediately from statement (1).

For statement (3), fix $i \in C$ and let $\xi^n = \underline{x}_i^{n+1}$. Then $\underline{x}_{-i}^n \to \underline{x}_{-i}$, $\xi^n \to \underline{x}_i$, and for every n, $\xi^n \in B_i(\underline{x}_{-i}^n)$. As B_i is upper hemicontinuous, it follows that $\underline{x}_i \in B_i(\underline{x}_{-i})$. Similarly, it can be shown that $\overline{x}_i \in B_i(\overline{x}_{-i})$.

For statement (4), fix $i \in S$ and let $\xi^n = \underline{x}_i^{n+1}$. Then $\overline{x}_{-i}^n \to \overline{x}_{-i}$, $\xi^n \to \underline{x}_i$, and for every n, $\xi^n \in B_i(\overline{x}_{-i}^n)$. As B_i is upper hemicontinuous, it follows that $\underline{x}_i \in B_i(\overline{x}_{-i})$. Similarly, $\overline{x}_i \in B_i(\underline{x}_{-i})$. □

Statements (3) and (4) in this theorem show that limits of lower and upper response dynamics are best responses to each other, with a change in where the best response is evaluated depending on type of player. As GSS is a special case of monotone game, these limits may not necessarily be Nash equilibria, but they do have behavioral content in terms of rationalizable strategies in the following sense.

In a monotone game, a profile of actions $x \in X$ is ***simply rationalizable***, if there is $y \in X$ such that for every player $i \in C$, $x_i \in B_i(x_{-i})$ and $y_i \in B_i(y_{-i})$, and for every player $i \in S$, $x_i \in B_i(y_{-i})$ and $y_i \in B_i(x_{-i})$. In this case, we say that x is simply rationalizable using y. Symmetry in the definition means that x is simply rationalizable using y, if, and only if, y is simply rationalizable using x. Moreover, when $C = \emptyset$, the definition specializes to that of simply rationalizable profile in a GSS.

Intuitively, a profile of actions x is simply rationalizable, if there is a profile of actions y such that every player i can rationalize playing x_i with a short cycle of conjectures using y as follows. A strategic complements player i plays x_i, because they believe their opponents will play x_{-i}, because each opponent j with strategic complements believes that others are playing x_{-j} and each opponent j with strategic substitutes believes that others are playing y_{-j}. Similarly, a strategic substitutes player i plays x_i, because they believe their opponents will play y_{-i}, because each opponent j with strategic complements believes that others are playing y_{-j} and each opponent j with strategic substitutes believes that others are playing x_{-j}.

As earlier, it is immediate from the definition that a Nash equilibrium x is simply rationalizable, because we may choose $y = x$. Simply rationalizable actions allow for additional outcomes as long as they are justifiable with short cycles of reasoning. This may be viewed as a type of bounded rationality. Simply rationalizable profiles are serially undominated.

Theorem 5.6 *If x is simply rationalizable using y, then both x and y are serially undominated.*

Proof Using induction, it is trivially true that $x, y \in \mathcal{U}^0$. Suppose $x, y \in \mathcal{U}^n$. Then for every $i \in C$, $x_i \in B_i(x_{-i})$ and $y_i \in B_i(y_{-i})$, and therefore, for every $i \in C$, $x_i \in \mathcal{U}_i^{n+1}$ and $y_i \in \mathcal{U}_i^{n+1}$. Similarly, for every $i \in S$, $x_i \in B_i(y_{-i})$ and $y_i \in B_i(x_{-i})$, and therefore, for every $i \in S$, $x_i \in \mathcal{U}_i^{n+1}$ and $y_i \in \mathcal{U}_i^{n+1}$. It follows that $x, y \in \mathcal{U}^{n+1}$, and therefore, $x, y \in \mathcal{U}^\infty$. □

Applying the definition to limits of lower and upper response dynamics, it follows immediately that both $\underline{x}$ and $\overline{x}$ are simply rationalizable using each other, and therefore, both are serially undominated. Both lower and upper response dynamics provide a constructive algorithm to compute these limits. In particular, with finitely many actions, both dynamics necessarily converge in finitely many iterations.

5.2.3 *Nash Equilibrium Set*

As monotone games are lattice games, Theorem 4.21 (page 101) shows that if one player has strict strategic substitutes and singleton valued best response, or if two players have strict strategic substitutes, then the Nash equilibrium set is totally unordered. This remains true for monotone games. The next result, due to Monaco and Sabarwal (2016), shows that in a lattice game if one player has strict strategic substitutes and one player has strict strategic complements, then again the Nash equilibrium set is totally unordered.

In a lattice game, player i has ***strict strategic complements***, if for every $\hat{x}_{-i}$ and $\tilde{x}_{-i}$ in X_{-i}, $\hat{x}_{-i} \prec \tilde{x}_{-i}$ implies $B_i(\hat{x}_{-i}) \sqsubset_c B_i(\tilde{x}_{-i})$. In other words, player i has strict strategic complements, if B_i is increasing in the completely lower than set order.

Theorem 5.7 *In a lattice game, if one player has strict strategic substitutes and one player has strict strategic complements, then the Nash equilibrium set is totally unordered.*

Proof Suppose, without loss of generality, player 1 has strict strategic substitutes and player 2 has strict strategic complements. Consider two distinct Nash equilibria $\hat{x}$ and $\tilde{x}$ and suppose $\hat{x}$ and $\tilde{x}$ are comparable, with $\hat{x} \prec \tilde{x}$. As case 1, suppose $\hat{x}_{-1} \prec \tilde{x}_{-1}$. Then $B_1(\tilde{x}_{-1}) \sqsubset_c B_1(\hat{x}_{-1})$. As $\tilde{x}_1 \in B_1(\tilde{x}_{-1})$ and $\hat{x} \in B_1(\tilde{x}_{-1})$, it follows that $\tilde{x}_1 \prec \hat{x}_1$, and this contradicts $\hat{x} \prec \tilde{x}$. As case 2, suppose $\hat{x}_{-1} = \tilde{x}_{-1}$ and $\hat{x}_1 \prec \tilde{x}_1$. Then $\hat{x}_{-2} \prec \tilde{x}_{-2}$, and therefore, $B_2(\hat{x}_{-2}) \sqsubset_c B_2(\tilde{x}_{-2})$. This implies that $\hat{x}_2 \prec \tilde{x}_2$ contradicting $\hat{x}_{-1} = \tilde{x}_{-1}$. □

Theorem 5.7 (page 141) and Theorem 4.21 (page 101) give several minimal conditions in a lattice game for the Nash equilibrium set to be totally unordered. If either (1) one player has strict strategic substitutes and singleton valued best response, or (2) two players have strict strategic substitutes, or (3) one player has strict strategic substitutes and one player has strict strategic complements, then the Nash equilibrium set is totally unordered. These theorems require no assumptions on payoffs of other players. Other players may have strategic complements, or strategic substitutes, or neither, and in each case, the equilibrium set is totally unordered.

The result about uniqueness of symmetric Nash equilibrium on chains, corollary 4.24 (page 104), remains true here without modification.

Examples of Nash equilibrium set for GSC or GSS are given in earlier chapters. Additional examples in which both types of players are present are as follows.

Examples 5.8 Consider the following examples.

1. In matching pennies (Example 1.3, page 4), the Nash equilibrium set is empty.
2. In online content provision, (Example 5.1, page 135), the Nash equilibrium set is $\mathcal{E} = \{(ML, ML), (ML, MH)\}$. It is a nonempty complete lattice. Indeed, it is a chain. In this example, player 1 has neither singleton valued best response nor strict strategic substitutes. Moreover, neither Nash equilibrium is strict.

3. In Cournot duopoly with spillovers (Example 4.22, page 100), firm 1 has strict strategic substitutes and singleton valued best response. Firm 2 has neither strategic complements nor strategic substitutes. The Nash equilibrium set is $\{(3.55, 32.90), (8.45, 23.10), (20, 0)\}$. It is totally unordered.
4. In Bernot duopoly (Example 5.2, page 135), the unique Nash equilibrium is given by $p_1^* = \frac{1}{13}(5a + 8c)$ and $q_2^* = \frac{3}{13}(a - c)$. For completeness, substituting these values in inverse market demands yields $q_1^* = \frac{5}{26}(a - c)$ and $p_2^* = \frac{1}{26}(9a + 17c)$. It follows that for reasonable values of a and c, firm 1 competing in price sets a higher price and produces less than firm 2 competing in quantity. For example, with $a = 270$ and $c = 10$, $p_1^* = 110$, $q_1^* = 50$, $p_2^* = 100$, $q_2^* = 60$.

5.2.4 *Dominance Solvability and Global Stability*

Development of bounds on undominated responses at the individual level for a strategic complements player (in Sect. 3.2.4, page 67) and for a strategic substitutes player (in Sect. 4.2.4, page 105) immediately imply the corresponding results for monotone games with both types of players. The same notation is used here as in Sect. 3.2.4 (page 67).

Theorem 5.9 *In every monotone game,*

1. *For every $i \in C$ and for every $x_{-i}, y_{-i} \in X_{-i}$ with $x_{-i} \preceq y_{-i}$,*

$$[\mathcal{U}_i([x_{-i}, y_{-i}])] = [\underline{B}_i(x_{-i}), \overline{B}_i(y_{-i})].$$

2. *For every $i \in S$ and for every $x_{-i}, y_{-i} \in X_{-i}$ with $x_{-i} \preceq y_{-i}$,*

$$[\mathcal{U}_i([x_{-i}, y_{-i}])] = [\underline{B}_i(y_{-i}), \overline{B}_i(x_{-i})].$$

3. *For every $x, y \in X$ with $x \preceq y$,*

$$[\mathcal{U}([x, y])] = [(\underline{B}_C(x), \underline{B}_S(y)), (\overline{B}_C(y), \overline{B}_S(x))].$$

Proof Statement (1) follows from the corresponding statement in Theorem 3.22 (page 68). Statement (2) follows from the corresponding statement in Theorem 4.26 (page 105). For statement (3), for $x, y \in X$ with $x \preceq y$,

$$\begin{aligned}[\mathcal{U}([x,y])] &= (\times_{i\in C}[\mathcal{U}_i([x_{-i},y_{-i}])]) \times (\times_{i\in S}[\mathcal{U}_i([x_{-i},y_{-i}])]) \\ &= \left(\times_{i\in C}[\underline{B}_i(y_{-i}),\overline{B}_i(x_{-i})]\right) \times \left(\times_{i\in S}[\underline{B}_i(y_{-i}),\overline{B}_i(x_{-i})]\right) \\ &= [(\underline{B}_C(x),\underline{B}_S(y)),(\overline{B}_C(y),\overline{B}_S(x))].\end{aligned}$$

□

Combined with earlier results on behavior of best response dynamics, the following results due to Barthel and Hoffmann (2019) follow in a straightforward manner. The continuity properties on payoffs here are weaker than the ones they used and upper hemicontinuity follows from the more general result for lattice games. The definition of adaptive dynamic and global stability is the same as earlier.

Theorem 5.10 *In a monotone game, let $(\underline{x}^n)_{n=1}^{\infty}$ be the lower response dynamic, $(\overline{x}^n)_{n=1}^{\infty}$ be the upper response dynamic, and $\underline{x}$ and $\overline{x}$ be their respective limit.*

1. *Every profile of serially undominated strategies is in the interval $[\underline{x},\overline{x}]$.*
2. *$\underline{x}$ is the smallest profile of serially undominated strategies in the game, and $\overline{x}$ is the largest profile of serially undominated strategies in the game.*
3. *$\underline{x}$ is the smallest simply rationalizable profile and $\overline{x}$ is the largest simply rationalizable profile.*
4. *Every adaptive dynamic $(x^n)_{n=0}^{\infty}$ satisfies $\underline{x} \preceq \liminf x^k \preceq \limsup x^k \preceq \overline{x}$.*

Proof For statement (1), Theorem 5.9 (page 142) and definitions of lower and upper response dynamics imply that for every $n \geq 0$,

$$[\mathcal{U}([\underline{x}^n,\overline{x}^n])] = [(\underline{B}_C(\underline{x}^n),\underline{B}_S(\overline{x}^n)),(\overline{B}_C(\overline{x}^n),\overline{B}_S(\underline{x}^n))] = [\underline{x}^{n+1},\overline{x}^{n+1}].$$

Therefore, the set of serially undominated strategies satisfies

$$\mathcal{U}^{\infty} \subset \bigcap_{n=0}^{\infty}[\mathcal{U}([\underline{x}^n,\overline{x}^n])] = \bigcap_{n=0}^{\infty}[\underline{x}^{n+1},\overline{x}^{n+1}] = [\underline{x},\overline{x}].$$

Statement (2) follows, because both $\underline{x}$ and $\overline{x}$ are serially undominated, and this is proved using induction on n. It is trivially true that $\underline{x},\overline{x} \in \mathcal{U}^0$. Suppose $\underline{x},\overline{x} \in \mathcal{U}^n$. Then for every $i \in C$, $\underline{x}_i \in B_i(\underline{x}_{-i})$ and $\overline{x}_i \in B_i(\overline{x}_{-i})$, and for every $i \in S$, $\underline{x}_i \in B_i(\overline{x}_{-i})$ and $\overline{x}_i \in B_i(\underline{x}_{-i})$. Therefore, for every i,

$\underline{x}_i \in \mathcal{U}_i^{n+1}$ and for every i, $\overline{x}_i \in \mathcal{U}_i^{n+1}$, whence $\underline{x}, \overline{x} \in \mathcal{U}^{n+1}$. It follows that both $\underline{x}$ and $\overline{x}$ are serially undominated. The conclusion now follows, because statement (1) shows that serially undominated strategies are bounded below by $\underline{x}$ and above by $\overline{x}$.

Statement (3) holds, because $\underline{x}$ and $\overline{x}$ are both simply rationalizable and every simply rationalizable profile is serially undominated.

Statement (4) is proved by first proving that for every $n \geq 0$, there is $k_n > n$, for every $k \geq k_n$, $x^k \in [\underline{x}_n, \overline{x}_n]$. This is proved using induction on n. This is trivial for $n = 0$, because $[\underline{x}^0, \overline{x}^0] = X$. Suppose the statement is true for fixed n. In this case, let $k_n > n$ be such that for every $k \geq k_n$, $x^k \in [\underline{x}^n, \overline{x}^n]$. Then for every $k > k_n$, $P(k_n, k) \subset [\underline{x}^n, \overline{x}^n]$. Moreover, applying the definition of adaptive dynamics to index k_n, let $k_{n+1} > k_n$ be such that for every $k \geq k_{n+1}$, $x^k \in [\mathcal{U}([P(k_n, k)])]$. Notice that for every $k \geq k_{n+1}$, $P(k_n, k) \subset [\underline{x}^n, \overline{x}^n]$. Therefore, $k_{n+1} > n + 1$, and for every $k \geq k_{n+1}$,

$$x^k \in [\mathcal{U}([P(k_n, k)])] \subset [\mathcal{U}([\underline{x}^n, \overline{x}^n])] = [\underline{x}^{n+1}, \overline{x}^{n+1}].$$

Statement (4) now follows, because $\underline{x} = \lim_n \underline{x}^n$ and $\overline{x} = \lim_n \overline{x}^n$. □

In the following theorem, statement (3) for unique simply rationalizable profile is new and not available in the previous literature.

Theorem 5.11 *In every monotone game, the following are equivalent.*

1. *The game is globally stable.*
2. *The game is dominance solvable.*
3. *The game has a unique simply rationalizable profile.*
4. $\underline{x} = \overline{x}$

Proof The equivalence of statements (2) and (4) follows from statements (1) and (2) of Theorem 5.10 (page 143). The equivalence of statements (3) and (4) follows from statement (3) of Theorem 5.10 (page 143). Statement (4) implies statement (1), using statement (4) of Theorem 5.10 (page 143). To see that statement (1) implies statement (4), notice that the sequence whose even terms consist of the lower response dynamic and whose odd terms consist of the upper response dynamic is an adaptive dynamic, and its convergence implies that both lower and upper response dynamics converge to the same limit. Therefore, $\underline{x} = \overline{x}$. □

Theorem 5.12 *In a monotone game, suppose the Nash equilibrium set is totally unordered. Let $\underline{x}$ be the limit of the lower response dynamic and $\overline{x}$ be the limit of the upper response dynamic.*

1. *$\underline{x}$ is a Nash equilibrium, if, and only if, $\underline{x}$ is the unique Nash equilibrium in the game.*
2. *$\overline{x}$ is a Nash equilibrium, if, and only if, $\overline{x}$ is the unique Nash equilibrium in the game.*
3. *With multiple Nash equilibria, neither $\underline{x}$ nor $\overline{x}$ are Nash equilibria.*

Proof The proof is the same as for Theorem 4.29 (page 109), using Theorem 5.10 (page 143) for bounds on the set of serially undominated strategies. □

This result shows that in monotone games, the behaviorally different notions of dominance solvability and global stability coincide. This is similar to the case for GSC and GSS. A difference is that uniqueness of Nash equilibrium is not necessarily sufficient for this conclusion, because it does not hold generally in GSS and GSS is a special case of monotone game.

This theorem provides computational algorithms to check for uniqueness of Nash equilibrium in monotone games. The algorithms given by the lower response dynamic and the upper response dynamic are both convergent algorithms. In finite games, each converges in finitely many iterations. Checking if the limit is a Nash equilibrium gives an algorithmic test for uniqueness.

5.2.5 *Monotone Comparative Statics of Equilibrium*

The effect of a change in the decision-making environment on predicted outcomes in situations where all players have strategic complements or all players have strategic substitutes is studied in earlier chapters. When both types of players are present simultaneously, the effect of a change in the decision-making environment on predicted outcomes is more complex. These effects are studied using parameterized monotone games, which are a subclass of parameterized lattice games.

A ***parameterized monotone game*** is a parameterized lattice game $\Gamma = ((X_i, u_i)_{i=1}^I, T)$, where

1. There are finitely many players, I, indexed by $i \in \{1, \ldots, I\}$.
2. For each i, the action space of player i is X_i. It is assumed to be a nonempty, subcomplete lattice in $\mathbb{R}^{n_i}$ with the Euclidean order. The space of profiles of actions is $X = X_1 \times \cdots \times X_I$ with the product order. The parameter space is $T \subset \mathbb{R}^n$. It is a poset with the Euclidean partial order.
3. For each i, the payoff for player i is $u_i : X_i \times X_{-i} \times T \to \mathbb{R}$, where

   (a) u_i is upper semicontinuous on X and continuous on X_{-i}, for every t,
   (b) For each (x_{-i}, t), $u_i(\cdot, x_{-i}, t)$ is quasisupermodular on X_i, and either u_i has single crossing property in (x_i, x_{-i}) (for each t), or u_i has dual single crossing property in (x_i, x_{-i}) (for each t), and
   (c) u_i has single crossing property in (x_i, t) (for each x_{-i}).

In other words, a parameterized monotone game is a parameterized lattice game in which for every i, either player i has strategic substitutes at every t, or player i has strategic complements at every t. Moreover, single crossing property in (x_i, t) implies that every player i has an incentive to take a higher action in t.

It follows immediately from the definition that in a parameterized monotone game, for every $t \in T$, the game at t, $\Gamma(t)$, is a monotone game.

In particular, for every strategic complements player $i \in C$, the section of best response correspondence for player i determined by t, $B_i(\cdot, t)$, mapping (x_{-i}, t) to $B_i(x_{-i}, t)$ is nonempty valued, compact valued, subcomplete valued, upper hemicontinuous in x_{-i}, and (weakly) increasing in x_{-i}, and for every strategic substitutes player $i \in S$, for every t, $B_i(\cdot, t)$ is nonempty valued, compact valued, subcomplete valued, upper hemicontinuous in x_{-i}, and (weakly) decreasing in x_{-i}. Moreover, single crossing property in (x_i, t) shows that for each x_{-i}, the section of best response correspondence for player i determined by x_{-i}, $B_i(x_{-i}, \cdot)$, mapping (x_{-i}, t) to $B_i(x_{-i}, t)$ is (weakly) increasing in t.

It follows that for each t, the joint best response correspondence of strategic complements players at t, $B_C(\cdot, t)$, mapping (x, t) to $B_C(x, t)$ is nonempty valued, compact valued, subcomplete valued, upper hemicontinuous in x, and (weakly) increasing in x, and the joint best response correspondence of strategic substitutes players at t, $B_S(\cdot, t)$, mapping (x, t) to $B_S(x, t)$ is nonempty valued, compact valued, subcomplete valued, upper hemicontinuous in x, and (weakly) decreasing in x. Moreover, for each x, the section of the overall joint best response correspondence determined by x, $B(x, \cdot) = (B_C(x, \cdot), B_S(x, \cdot))$, mapping (x, t) to $B(x, t)$ is (weakly) increasing in t.

The equilibrium correspondence is $\mathcal{E} : T \rightrightarrows X$, mapping t to $\mathcal{E}(t)$.

Earlier chapters include results that are stated at the level of parameterized lattice game. As parameterized monotone game is a special case, those results remain true here. Some of those results are as follows.

In a parameterized lattice game $\Gamma = ((X_i, u_i)_{i=1}^I, T)$, for each $t \in T$, the game at t, $\Gamma(t)$, is a lattice game and therefore, with conditions on one or two players only, the equilibrium set at t, $\mathcal{E}(t)$ is totally unordered (see Theorem 4.21 (page 101) or Theorem 5.7 (page 141)).

When players have incentive to take higher action in t, under mild conditions, the equilibrium correspondence in a parameterized lattice game is never decreasing. Theorem 4.35 (page 115) provides conditions on one or two players only to conclude that the equilibrium correspondence is never decreasing. There are no additional restrictions on payoffs for other players at other parameter values.

In these situations, when action spaces are chains, an increase in the parameter implies that equilibrium action of at least one player goes up and therefore, the profile of actions in symmetric equilibrium goes up. Theorem 4.36 (page 117) shows that in a parameterized lattice game in which action space of each player is a chain, a never decreasing equilibrium correspondence implies that whenever the parameter increases, at least one player takes a higher equilibrium action, and therefore, if both the old and the new equilibrium are symmetric, the new equilibrium is necessarily higher.

When symmetric equilibrium is unique, it follows that an increase in the parameter leads to an increase in the symmetric equilibrium. Theorem 4.37 (page 118) shows that under mild conditions on one or two players only, the symmetric Nash equilibrium correspondence is singleton valued and (weakly) increasing.

Beyond these results, an increase in the parameter does not necessarily lead to an increase in the profile of equilibrium actions, even in two player monotone games in which action spaces are chains, best responses are singleton valued, and Nash equilibrium is unique.

Example 5.13 (Parameterized Bernot duopoly) Consider two firms producing branded handbags in Bernot duopoly (Example 5.2, page 135).

Firm 1 focuses on selling its product by choosing price and firm 2 focuses on selling its product by choosing quantity of product to sell. Inverse market demand for firm 1 is given by $p_1 = a - 2q_1 - q_2$ and for firm 2 by $p_2 = a - 2q_2 - q_1$, where p_1 is price of firm 1 handbags, p_2 is price of firm 2 handbags, q_1 is output of firm 1 handbags, q_2 is output of firm 2 handbags, and $a > 0$ is an index for strength of demand. Handbags are produced at constant marginal cost $c > 0$. Firm 1 chooses p_1 and firm 2 chooses q_2. Profit of each firm is given by total revenue minus total cost.

Consider a situation in which firm 1 benefits from an increase in intensity of preference for its product and firm 2 benefits from technological improvement to reduce its cost of production. For convenience both are summarized by an innovation parameter $t \geq 0$. Moreover, suppose impact of these innovations may be asymmetric for the two firms. For convenience, this asymmetry is summarized by relative intensity of innovation adoption $\alpha \in [0, 1]$. With this notation, inverse market demand for firm 1 is given by $p_1 = a + (1 - \alpha)t - 2q_1 - q_2$ and constant marginal cost for firm 2 is given by $c - \alpha t > 0$. Intuitively, for fixed innovation t, α closer to 0 indicates relatively high impact on intensity of preference for firm 1 handbags and relatively low impact on cost savings for firm 2, and α closer to 1 indicates the opposite.

Profit of firm 1 is given by $\pi_1(p_1, q_2, t) = (p_1 - c)\frac{1}{2}(a + (1 - \alpha)t - q_2 - p_1)$. This function satisfies dual single crossing property in (p_1, q_2) for each t, because $\frac{\partial^2 \pi_1}{\partial q_2 \partial p_1} = -\frac{1}{2} < 0$, and this function satisfies single crossing property in (p_1, t) for each q_2, because $\frac{\partial^2 \pi_1}{\partial t \partial p_1} = \frac{1-\alpha}{2} \geq 0$. Therefore, firm 1 has strategic substitutes at every t and has an incentive to take higher action in t.

Profit of firm 2 is given by $\pi_2(p_1, q_2, t) = (a - \frac{1}{2}(a + (1 - \alpha)t - q_2 - p_1) - 2q_2 - c + \alpha t)q_2$. This function satisfies single crossing property in (p_1, q_2) for each t, because $\frac{\partial^2 \pi_2}{\partial q_2 \partial p_1} = \frac{1}{2} > 0$, and for $\alpha \geq \frac{1}{3}$, this function satisfies single crossing property in (q_2, t) for each p_1, because $\frac{\partial^2 \pi_1}{\partial t \partial q_2} = \frac{3\alpha - 1}{2} \geq 0$ for $\alpha \geq \frac{1}{3}$. Therefore, in the remainder of this example, suppose $\alpha \geq \frac{1}{3}$. Firm 2 has strategic complements at every t and has an incentive to take higher action in t.

It can be calculated that best response of firm 1 is $B_1(q_2) = \frac{1}{2}(a + c + (1 - \alpha)t - q_2)$ a function that is decreasing in q_2 and increasing in t, and best response of firm 2 is $B_2(p_1) = \frac{1}{6}(a - 2c + (3\alpha - 1)t + p_1)$ a function that is increasing in p_1 and increasing in t. The unique Nash equilibrium is given by

$$p_1^* = \frac{1}{13}(5a + 8c + (7 - 9\alpha)t) \quad \text{and} \quad q_2^* = \frac{1}{78}(18a - 18c + (30\alpha - 6)t).$$

It follows that q_2^* is increasing in t, and p_1^* is increasing in t, if, and only if $\alpha < \frac{7}{9}$. For $\frac{7}{9} < \alpha$, p_1^* decreases when t goes up.

In other words, when innovation impact is sufficiently disproportionately larger for firm 2, firm 1 price may go down with an increase in intensity of demand for its own product.

Violation of monotone comparative statics of equilibrium outcomes in the two player case may be viewed in terms of direct and indirect effects. If the impact of an increase in parameter is large for firm 2 (the one with strategic complements) then the indirect strategic substitute effect may be large for firm 1 (the one with strategic substitutes). In addition, if the direct parameter effect is small for firm 1 it may be insufficient to overcome the large indirect strategic substitute effect causing equilibrium action of firm 1 to go down. On the other hand, as shown in the result below due to Monaco and Sabarwal (2016), if the combined effect is favorable, monotone comparative statics of equilibrium outcomes is obtained.

Theorem 5.14 (MCS-Two players) *Consider a two player parameterized monotone game in which player 1 has strategic substitutes and player 2 has strategic complements. Suppose each player's action space is a chain and best response is singleton valued.*

For each $\hat{t}, \tilde{t} \in T$ with $\hat{t} \prec \tilde{t}$, and for each $x^* \in \mathcal{E}(\hat{t})$ and $y^* \in \mathcal{E}(\tilde{t})$,

1. $x_2^* \preceq y_2^*$
2. $x_1^* \preceq y_1^*$, *if, and only if,* $x_1^* \preceq B_1(B_2(x_1^*, \tilde{t}), \tilde{t})$

Proof For statement (1), suppose first that $x_1^* \preceq y_1^*$. In this case $x_2^* = B_2(x_1^*, \hat{t}) \preceq B_2(y_1^*, \hat{t}) \preceq B_2(y_1^*, \tilde{t}) = y_2^*$, where the first inequality follows from the assumption that player 2 has strategic complements and the second from player 2 payoff having single crossing property in (x_2, t). Otherwise, $y_1^* \prec x_1^*$, and in this case if $y_2^* \prec x_2^*$, then $x_1^* = B_1(x_2^*, \hat{t}) \preceq B_1(x_2^*, \tilde{t}) \preceq B_1(y_2^*, \tilde{t}) = y_1^*$, where the first inequality follows from the assumption that player 1 payoff has single crossing property in (x_1, t) and the second from player 1 having strategic substitutes. This contradicts $y_1^* \prec x_1^*$. Therefore, $x_2^* \preceq y_2^*$.

For statement (2), if $x_1^* \preceq y_1^*$, then player 2 has strategic complements implies $B_2(x_1^*, \tilde{t}) \preceq B_2(y_1^*, \tilde{t})$ and taken together these imply $x_1^* \preceq y_1^* \preceq B_1(B_2(y_1^*, \tilde{t}), \tilde{t}) \preceq B_1(B_2(x_1^*, \tilde{t}), \tilde{t})$, where the last inequality follows from the assumption that player 1 has strategic substitutes. On the other hand, if $y_1^* \prec x_1^*$, then player 2 has strategic complements implies $y_2^* = B_2(y_1^*, \tilde{t}) \preceq B_2(x_1^*, \tilde{t})$ and taken together these imply a $B_1(B_2(x_1^*, \tilde{t}), \tilde{t}) \preceq B_1(y_2^*, \tilde{t}) = y_1^* \prec x_1^*$, where the weak inequality follows from the assumption that player 1 has strategic substitutes. □

Under the assumptions of this theorem, in two player parameterized monotone games, the equilibrium action of the player with strategic complements always goes up, regardless of the equilibria being compared. The equilibrium action of the player with strategic substitutes goes up as long as the combined direct and indirect effect is favorable, in the sense that $x_1^* \preceq B_1(B_2(x_1^*, \tilde{t}), \tilde{t})$. The condition $x_1^* \preceq B_1(B_2(x_1^*, \tilde{t}), \tilde{t})$ does not depend on the new equilibrium, and therefore, when this condition is satisfied, every equilibrium y^* at $\tilde{t}$ is higher than x^*. This gives a uniform monotone comparative statics result.

The condition $x_1^* \preceq B_1(B_2(x_1^*, \tilde{t}), \tilde{t})$ for the strategic substitutes player has the same form as the sufficient condition used in parameterized GSS (see Theorem 4.39, page 124). A difference is that it is necessary and sufficient here. Moreover, no condition is needed for the strategic complements player. This condition may be viewed graphically as follows.

In Fig. 5.1 (page 151), player 1 has strategic substitutes and player 2 has strategic complements. After an increase in parameter player 1 best response shifts out and player 2 best response shifts up. The composite effect for player 1 is favorable and correspondingly, equilibrium action of both players goes up.

In Fig. 5.2 (page 151), the composite effect for player 1 is unfavorable and correspondingly, equilibrium action of player 1 goes down. Equilibrium action of player 1 goes down precisely because the direct parameter effect for player 1 is small (as given by a small shift out of best response) and is insufficient to overcome the large opposite indirect strategic substitute effect (given by a large shift out of best response of player 2). Equilibrium action of player 2 goes up.

When action spaces are not chains, both statements in Theorem 5.14 (page 149) do not necessarily generalize.

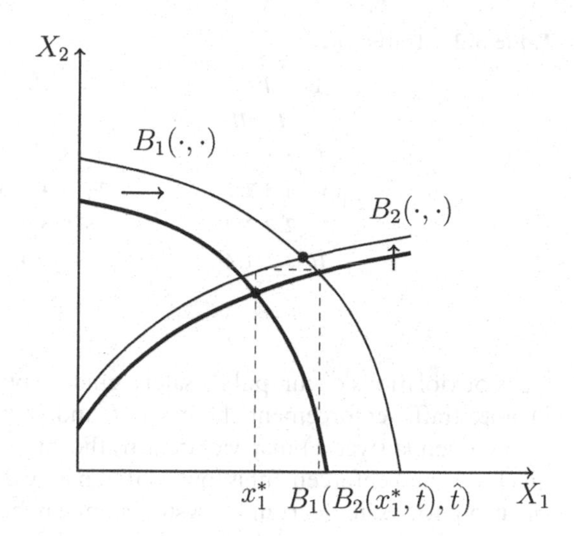

Fig. 5.1 Favorable composition of direct and indirect effects

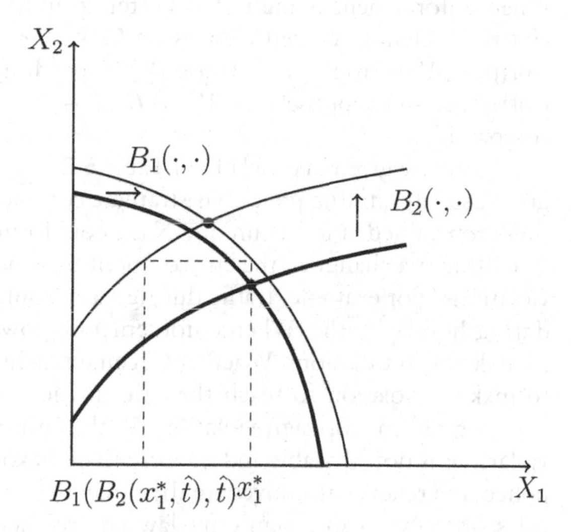

Fig. 5.2 Unfavorable composition of direct and indirect effects

Example 5.15 (Traffic safety) Consider traffic enforcement with two types of traffic violations, one for running a stop sign, another for speeding.

Traffic police (player 2) can choose a level of enforcement that is low or high. When citizens respect the rules or violate a stop sign only, law enforcement may remain low. When speeding is involved or when both

Table 5.2 Traffic safety

(1) P2

P1	L	H
L	3, 1	3, 0
V_1	4, 3	2, 2
V_2	2, 2	2, 3
H	3, 3	1, 4

(2) P2

P1	L	H
L	3, 2	4, 1
V_1	5, 4	3, 3
V_2	8, 3	7, 5
H	9, 4	4, 7

types of violations occur, public safety guides law enforcement to be high. Denote traffic enforcement choices as L and H with $L \prec H$.

A citizen (player 1) may violate a traffic rule to reach their destination quickly. When enforcement is low, a citizen may decide to make one type of violation, for example, running a stop sign may be preferred over speeding. When enforcement is high, it is better to follow the rules and make no violation. Denote citizen choices as L, V_1, V_2, H, where L signifies no violation, V_1 is stop sign violation, V_2 is speeding violation, and H is both violations, and suppose $L \prec V_1 \prec H$, $L \prec V_2 \prec H$, and V_1, V_2 are not comparable.

Payoffs are given in panel (1) of Table 5.2 (page 152). Citizen has strategic substitutes, traffic police has strategic complements, best responses are singleton valued, and the unique Nash equilibrium is (V_1, L).

Consider a change in the environment that increases incidence of traffic violations. For example, traffic during rush hour, or on Saturday night, or during holiday traffic. When enforcement is low, a citizen is more willing to make both violations. When enforcement is high, a citizen is still willing to make a violation to reach their destination quickly, and say, speeding is preferable to stop sign violation. With high enforcement, making two violations is not advisable and is worse than making no violation. For traffic police, the relative incentives are the same, that is, when citizens respect the rules or violate a stop sign only, law enforcement remains low, but when speeding is involved or when both types of violations occur, it is better to have high law enforcement. Traffic police payoffs are higher in these cases. This is summarized by payoffs in panel (2) of Table 5.2 (page 152). This remains a monotone game and the unique Nash equilibrium is (V_2, H).

If we denote the first environment as initial situation (with parameter $\hat{t} = 0$) and the second environment with greater incentives for traffic viola-

Table 5.3 Traffic safety, extended

(1)

		P2			
		L	ML	MH	H
	L	1, 5	5, 3	4, 2	5, 1
P1	V_1	3, 2	5, 3	4, 4	3, 3
	V_2	3, 3	3, 4	3, 3	2, 2
	H	4, 2	3, 3	3, 4	1, 5

(2)

		P2			
		L	ML	MH	H
	L	1, 5	6, 3	5, 2	7, 1
P1	V_1	6, 3	6, 4	5, 5	5, 4
	V_2	6, 4	8, 5	7, 4	4, 3
	H	8, 3	8, 5	7, 6	2, 8

tion as a situation with a higher parameter ($\tilde{t} = 1$), this is a parameterized monotone game. The condition in statement (2) of Theorem 5.14 (page 149) is satisfied, because $V_1 \prec H = B_1(B_2(V_1, 1), 1)$, but V_2 in the new equilibrium is not higher than V_1 in the old equilibrium. Even though enforcement is higher in the new equilibrium, the new profile of equilibrium actions is not comparable to the old profile.

Example 5.16 (Traffic safety, extended) Consider traffic enforcement example 5.15 (page 151) with traffic police having four graduated levels of enforcement, L, ML, MH, and H with $L \prec ML \prec MH \prec H$. Suppose initial payoffs are as in panel (1) of Table 5.3 (page 153) and payoffs after parameter increase are as in panel (2). As in the earlier example, this is a parameterized monotone game. Unique Nash equilibrium before parameter change is (V_1, MH) and after parameter change is (V_2, ML). Equilibrium enforcement goes down from MH to ML. In other words, when action space is not a chain, action of the strategic complements player may go down with an increase in the parameter.

As earlier, the computation $x_1^* \preceq B_1(B_2(x_1^*, \tilde{t}), \tilde{t})$ is equivalent to $B_1(B_2(x_1^*, \hat{t}), \hat{t}) = x_1^* \preceq B_1(B_2(x_1^*, \tilde{t}), \tilde{t})$, that is, $B_1(B_2(x_1^*, t), t)$ is (weakly) increasing in t at $\hat{t}$. Similar to the case of parameterized GSS (Theorem 4.41, page 126), when payoff functions are twice continuously differentiable, the condition $B_1(B_2(x_1^*, t), t)$ is (weakly) increasing in t at $\hat{t}$ may be formulated in terms of an equivalent condition on payoff functions, making it more accessible and easier to apply. In the following theorem, for notational convenience, superscript is used to index payoff function of player i, denoted u^i, and subscripts to denote partial derivatives. That is, $u^i_{i,t} = \frac{\partial^2 u^i}{\partial t \partial x_i}$, $u^i_{i,j} = \frac{\partial^2 u^i}{\partial x_j \partial x_i}$, and so on.

Theorem 5.17 (MCS-Two player case, differentiable payoffs) *Consider a two player parameterized monotone game in which player 1 has strategic substitutes and player 2 has strategic complements. Suppose each player's action space is an interval in the reals, their payoff is twice continuously differentiable, and their payoff is strictly concave in own variable. For each $\hat{t}$ in the interior of T and $x^* \in \mathcal{E}(\hat{t})$, the following are equivalent.*

1. *The composition $B_1(B_2(x_1^*, t), t)$ is increasing in t at $\hat{t}$.*
2. *For player 1,*

$$\left[u^1_{1,t} + u^1_{1,2} \left(-\frac{u^2_{2,t}}{u^2_{2,2}} \right) \right] \Bigg|_{(x^*,\hat{t})} > 0.$$

Proof Similar to proof of Theorem 4.41 (page 126). □

Example 5.18 (Parameterized Bernot duopoly, continued) In the parameterized Bernot duopoly (Example 5.13, page 147), payoffs are given by

$$\pi^1(p_1, q_2, t) = (p_1 - c)\tfrac{1}{2}(a + (1-\alpha)t - q_2 - p_1)$$
$$\pi^2(p_1, q_2, t) = (a - \tfrac{1}{2}(a + (1-\alpha)t - q_2 - p_1) - 2q_2 - c + \alpha t)q_2.$$

In this case, $\pi^1_{1,t} = \frac{1}{2}(1-\alpha)$, $\pi^1_{1,2} = -\frac{1}{2}$, $\pi^2_{2,t} = \frac{1}{2}(3\alpha - 1)$, and $\pi^2_{2,2} = -3$. The condition in the theorem is $\frac{1}{2}(1-\alpha) - \frac{1}{2}(\frac{3\alpha-1}{6}) > 0$, which is satisfied exactly when $\alpha > \frac{7}{9}$.

More generally, the next two theorems give conditions under which, for a general parameterized monotone game, monotone comparative statics of equilibrium outcomes is obtained. These theorems are new results unavailable in the previous literature. They do not invoke a fixed point theorem, or convex action spaces, or convex valued best responses.

For $\hat{t}, \tilde{t} \in T$ with $\hat{t} \prec \tilde{t}$, ***player*** i ***has strictly higher actions at*** $(\hat{t}, \tilde{t})$, if for every x_{-i}, $B_i(x_{-i}, \hat{t}) \sqsubset_c B_i(x_{-i}, \tilde{t})$. In other words, player i has strictly higher actions at $\hat{t} \prec \tilde{t}$, if an increase from $\hat{t}$ to $\tilde{t}$ implies that $B_i(x_{-i}, \hat{t})$ increases to $B_i(x_{-i}, \tilde{t})$ in the completely lower than set order.

Theorem 5.19 (MCS-Parameterized monotone games) *In a parameterized monotone game, for each $\hat{t}, \tilde{t} \in T$ with $\hat{t} \prec \tilde{t}$, and for each $x^* \in \mathcal{E}(\hat{t})$, let $\tilde{y}$ be defined as follows. For $i \in S$, let $\tilde{y}_i = \overline{B}_i(x^*_{-i}, \tilde{t})$, and for $i \in C$, let*

$\tilde{y}_i = \overline{B}_i(\overline{x}_{C_{-i}}, \tilde{y}_S, \tilde{t})$, where $C_{-i} = C \setminus \{i\}$ and $\overline{x}_{C_{-i}} = \sup X_{C_{-i}}$. Let $\tilde{\Gamma}(\tilde{t})$ be the lattice game at $\tilde{t}$ restricted to $[x^*, \tilde{y}]$ and $\tilde{\mathcal{E}}(\tilde{t})$ be the Nash equilibrium set for $\tilde{\Gamma}(\tilde{t})$.

1. *For every $\hat{t} \prec \tilde{t}$, if $x_S^* \preceq \underline{B}_S(\tilde{y}, \tilde{t})$, then $\tilde{\mathcal{E}}(\tilde{t}) \subset \mathcal{E}(\tilde{t})$ and for every $y^* \in \tilde{\mathcal{E}}(\tilde{t})$, $x^* \preceq y^*$.*
2. *For every $\hat{t} \prec \tilde{t}$, if $x_S^* \prec \underline{B}_S(\tilde{y}, \tilde{t})$, then $\tilde{\mathcal{E}}(\tilde{t}) \subset \mathcal{E}(\tilde{t})$ and for every $y^* \in \tilde{\mathcal{E}}(\tilde{t})$, $x^* \prec y^*$.*
3. *For every $\hat{t} \prec \tilde{t}$, if for every $i \in S$, $x_i^* \prec \underline{B}_i(\tilde{y}_{-i}, \tilde{t})$, then $\tilde{\mathcal{E}}(\tilde{t}) \subset \mathcal{E}(\tilde{t})$ and for every $y^* \in \tilde{\mathcal{E}}(\tilde{t})$ and for every $i \in S$, $x_i^* \prec y_i^*$.*
4. *For every $\hat{t} \prec \tilde{t}$, if for every $i \in S$, $x_i^* \prec \underline{B}_i(\tilde{y}_{-i}, \tilde{t})$, and for every $i \in C$, player i has strictly higher actions at $(\hat{t}, \tilde{t})$, then $\tilde{\mathcal{E}}(\tilde{t}) \subset \mathcal{E}(\tilde{t})$ and for every $y^* \in \tilde{\mathcal{E}}(\tilde{t})$ and for every i, $x_i^* \prec y_i^*$.*

Proof Consider $\hat{t} \prec \tilde{t}$, $x^* \in \mathcal{E}(\hat{t})$, and let $\tilde{y}$ be as in the hypothesis. For each strategic substitutes player $i \in S$, single crossing property in (x_i, t) implies $x_i^* \preceq \overline{B}_i(x_{-i}^*, \tilde{t}) = \tilde{y}_i$. Moreover, for each strategic complements player $i \in C$, $x_i^* \preceq \overline{B}_i(x_{-i}^*, \tilde{t}) \preceq \overline{B}_i(\overline{x}_{C_{-i}}, \tilde{y}_S, \tilde{t}) = \tilde{y}_i$, where the first inequality follows from single crossing property in (x_i, t) and the second from strategic complements. Therefore, $x^* \preceq \tilde{y}$.

Consider the nonempty interval $[x^*, \tilde{y}]$, let $\tilde{\Gamma}(\tilde{t})$ be the lattice game at $\tilde{t}$ restricted to $[x^*, \tilde{y}]$, and let $\tilde{\mathcal{E}}(\tilde{t})$ be the Nash equilibrium set for $\tilde{\Gamma}(\tilde{t})$. If $\tilde{\mathcal{E}}(\tilde{t}) = \emptyset$, each of the four statements is vacuously true.

The hypothesis in statement (1) implies that for every $z \in [x^*, \tilde{y}]$, $B(z, \tilde{t}) \subset [x^*, \tilde{y}]$, and this is seen as follows. Consider $z \in [x^*, \tilde{y}]$. For each $i \in S$, $z_{-i} \preceq \tilde{y}_{-i}$ implies $B_i(\tilde{y}_{-i}, \tilde{t}) \sqsubseteq B_i(z_{-i}, \tilde{t})$, and combined with the hypothesis in statement (1), it follows that $x_i^* \preceq \underline{B}_i(\tilde{y}, \tilde{t}) \preceq \underline{B}_i(z, \tilde{t})$. Moreover, $x_{-i}^* \preceq z_{-i}$ implies $B_i(z_{-i}, \tilde{t}) \sqsubseteq B_i(x_{-i}^*, \tilde{t})$, and therefore, $\overline{B}_i(z_{-i}, \tilde{t}) \preceq \overline{B}_i(x_{-i}^*, \tilde{t}) = \tilde{y}_i$. It follows that $B_i(z_{-i}, \tilde{t}) \subset [x_i^*, \tilde{y}_i]$. For each $i \in C$, strict single crossing property in (x_i, t) implies $x_i^* \preceq \overline{B}_i(x_{-i}^*, \hat{t}) \preceq \underline{B}_i(x_{-i}^*, \tilde{t})$, and combined with strategic complements and $\tilde{y}_{-i} \preceq (\overline{x}_{C_{-i}}, \tilde{y}_S)$, it follows that $\overline{B}_i(\tilde{y}_{-i}, \tilde{t}) \preceq \overline{B}_i(\overline{x}_{C-i}, \tilde{y}_S, \tilde{t}) = \tilde{y}_i$. Therefore, $B_i(z_{-i}, \tilde{t}) \subset [x_i^*, \tilde{y}_i]$.

Consider the restricted game $\tilde{\Gamma}(\tilde{t})$. For every $z \in [x^*, \tilde{y}]$ and for every player i, let $\tilde{B}_i(z_{-i}, \tilde{t})$ denote the best response correspondence of player i in the restricted game, that is, $\tilde{B}_i(z_{-i}, \tilde{t}) = \arg\max_{\xi \in [x_i^*, \tilde{y}_i]} u_i(\xi, z_{-i}, \tilde{t})$. Notice that for every $z \in [x^*, \tilde{y}]$ and for every player i, $\tilde{B}_i(z_{-i}, \tilde{t}) = B_i(z_{-i}, \tilde{t})$, as follows. If $x_i \in B_i(z_{-i}, \tilde{t})$, then for every $\xi \in X_i$, $u_i(x_i, z_{-i}, \tilde{t}) \geq$

$u_i(\xi_i, z_{-i}, \tilde{t})$, and therefore, for every $\xi \in [x_i^*, \tilde{t}_i]$, $u_i(x_i, z_{-i}, \tilde{t}) \geq u_i(\xi_i, z_{-i}, \tilde{t})$, implying $\tilde{B}_i(z_{-i}, \tilde{t}) \subset B_i(z_{-i}, \tilde{t})$. In the other direction, if $x_i \in \tilde{B}_i(z_{-i}, \tilde{t})$, then for every $\xi \in [x_i^*, \tilde{t}_i]$, $u_i(x_i, z_{-i}, \tilde{t}) \geq u_i(\xi_i, z_{-i}, \tilde{t})$, and as $B_i(x_{-i}, \tilde{t}) \subset [x_i^*, \tilde{y}_i]$, the maximum is achieved in $[x_i^*, \tilde{y}_i]$, and therefore, for every $\xi \in X_i$, $u_i(x_i, z_{-i}, \tilde{t}) \geq u_i(\xi_i, z_{-i}, \tilde{t})$, implying $\tilde{B}_i(z_{-i}, \tilde{t}) \subset B_i(z_{-i}, \tilde{t})$.

Let $\tilde{B}(z, \tilde{t})$ denote the joint best response correspondence in the restricted game. The previous argument shows that for every $z \in [x^*, \tilde{y}]$, $\tilde{B}(z, \tilde{t}) = B(z, \tilde{t})$, and therefore, every fixed point of $\tilde{B}(\cdot, \tilde{t})$ is a fixed point of $B(\cdot, \tilde{t})$, and this implies $\tilde{\mathcal{E}}(\tilde{t}) \subset \mathcal{E}(\tilde{t})$. Consider $y^* \in \tilde{\mathcal{E}}(\tilde{t}) \subset [x^*, \tilde{y}]$. For each $i \in C$, strict single crossing property in (x_i, t) and strategic complements imply $x_i^* \preceq \overline{B}_i(x_{-i}^*, \hat{t}) \preceq \underline{B}_i(x_{-i}^*, \tilde{t}) \preceq \underline{B}_i(y_{-i}^*, \tilde{t}) \preceq y_i^*$. For each $i \in S$, the hypothesis in statement (1) and strategic substitutes imply that $x_i^* \preceq \underline{B}_i(\tilde{y}_{-i}, \tilde{t}) \preceq \underline{B}_i(y_{-i}^*, \tilde{t}) \preceq y_i^*$. Consequently, $x^* \preceq y^*$.

Statement (2) follows from the same argument and modifying the last sentence in the argument with the hypothesis in statement (2), that is, $x_S^* \prec \underline{B}_S(\tilde{y}, \tilde{t}) \preceq \underline{B}_S(y^*, \tilde{t}) \preceq y_S^*$. This implies $x^* \prec y^*$. Statement (3) follows from the same argument as well, modifying the last sentence in the argument with the hypothesis in statement (3), that is, for every $i \in S$, $x_i^* \prec \underline{B}_i(\tilde{y}_{-i}, \tilde{t}) \preceq \underline{B}_i(y_{-i}^*, \tilde{t}) \preceq y_i^*$. Statement (4) follows from an extension of statement (3) by noting that for every $i \in C$, player i has strictly higher actions at $(\hat{t}, \tilde{t})$ implies $x_i^* \preceq \overline{B}_i(x_{-i}^*, \hat{t}) \prec \underline{B}_i(x_{-i}^*, \tilde{t}) \preceq \underline{B}_i(y_{-i}^*, \tilde{t}) \preceq y_i^*$. □

The condition $x_S^* \preceq \underline{B}_S(\tilde{y}, \tilde{t})$ may be motivated in terms of direct and indirect effects. For convenience, suppose best response of each player is singleton valued.

Consider an equilibrium profile at the lower parameter value, $x^* \in \mathcal{E}(\hat{t})$. When the parameter increases to $\tilde{t}$, the direct parameter effect provides an incentive to each player to take a higher action. The joint direct effect for strategic complements players is $B_C(x^*, \tilde{t})$, for strategic substitutes players is $B_S(x^*, \tilde{t})$, and for all players is $B(x^*, \tilde{t})$.

The indirect effect for strategic complements players reinforces the direct effect and no additional restriction is needed for these players.

The indirect effect for strategic substitutes players goes in the opposite direction. For each player with strategic substitutes, this effect is $B_i(B(x^*, \tilde{t})_{-i}, \tilde{t})$. In parameterized GSS, if this composite effect is favorable, that is, $x_i^* \preceq B_i(B(x^*, \tilde{t})_{-i}, \tilde{t})$, it is a sufficient condition for monotone comparative statics. When there are players with strategic complements, this may no longer be sufficient in general, because the indirect strategic complements effect further increases the action of players with strategic complements and this implies a greater move in the opposite direction by players with strategic substitutes. As strength of strategic complements may vary, a sufficient condition would need to account for the strongest possible indirect strategic complements effect, and for each strategic complements player, this is given by $\tilde{y}_i = B_i(\overline{x}_{C_{-i}}, \tilde{y}_S, \tilde{t})$, with the largest profile $\overline{x}_{C_{-i}}$ for strategic complements players (other than i).

For each strategic substitutes player $i \in S$, if the composite effect given by $B_i(\tilde{y}, \tilde{t})$ is favorable, that is, $x_i^* \preceq B_i(\tilde{y}, \tilde{t})$, then monotone comparative statics of equilibrium outcomes is obtained. No additional conditions are needed for players with strategic complements.

Theorem 5.19 (page 154) is formulated without invoking applicability of a particular fixed point theorem or other device to guarantee existence of equilibrium. In a parameterized lattice game with Nash equilibrium on subintervals, it follows that $\tilde{\mathcal{E}}(\tilde{t}) \neq \emptyset$, and monotone comparative statics of equilibrium outcomes follows.

Theorem 5.20 (MCS-Parameterized monotone games, part 2) *In a parameterized monotone game, for each $\hat{t}, \tilde{t} \in T$ with $\hat{t} \prec \tilde{t}$, and for each $x^* \in \mathcal{E}(\hat{t})$, let $\tilde{y}$ be defined as follows. For $i \in S$, $\tilde{y}_i = \overline{B}(x^*_{-i}, \tilde{t})$, and for $i \in C$, let $\tilde{y}_i = \overline{B}(\overline{x}_{C_{-i}}, \tilde{y}_S, \tilde{t})$, where $C_{-i} = C \setminus \{i\}$ and $\overline{x}_{C_{-i}} = \sup X_{C_{-i}}$. Let $\tilde{\Gamma}(\tilde{t})$ be the lattice game at $\tilde{t}$ restricted to $[x^*, \tilde{y}]$, let $\tilde{\mathcal{E}}(\tilde{t})$ be the Nash equilibrium set for $\tilde{\Gamma}(\tilde{t})$, and suppose $\tilde{\mathcal{E}}(\tilde{t}) \neq \emptyset$.*

1. *For every $\hat{t} \prec \tilde{t}$, if $x_S^* \preceq \underline{B}_S(\tilde{y}, \tilde{t})$, then there is $y^* \in \mathcal{E}(\tilde{t})$ such that $x^* \preceq y^*$.*
2. *For every $\hat{t} \prec \tilde{t}$, if $x_S^* \prec \underline{B}_S(\tilde{y}, \tilde{t})$, then there is $y^* \in \mathcal{E}(\tilde{t})$ such that $x^* \prec y^*$.*
3. *For every $\hat{t} \prec \tilde{t}$, if for every $i \in S$, $x_i^* \prec \underline{B}_i(\tilde{y}_{-i}, \tilde{t})$, then there is $y^* \in \mathcal{E}(\tilde{t})$ such that for every $i \in S$, $x_i^* \prec y_i^*$.*
4. *For every $\hat{t} \prec \tilde{t}$, if for every $i \in S$, $x_i^* \prec \underline{B}_i(\tilde{y}_{-i}, \tilde{t})$ and for every $i \in C$, player i has strictly higher actions at $(\hat{t}, \tilde{t})$, then there is $y^* \in \mathcal{E}(\tilde{t})$ such that for every i, $x_i^* \prec y_i^*$.*

Proof As $\tilde{\mathcal{E}}(\tilde{t}) \neq \emptyset$, let $y^* \in \tilde{\mathcal{E}}(\tilde{t})$. By the previous theorem, $y^* \in \mathcal{E}(\tilde{t})$. Each statement follows by applying the corresponding statement in Theorem 5.19 (page 154) to this y^*. □

Theorem 5.20 (page 157) generalizes the result due to Monaco and Sabarwal (2016) for the special case of convex action spaces and convex valued best response correspondence.

As a whole, this book develops foundations of the theory of monotone games in a unified manner. Earlier chapters study games in which either all players have strategic complements (GSC) or all players have strategic substitutes (GSS). This chapter develops the theory of games in which both types of players are present simultaneously. Several results for GSS generalize to this class of games. New results blend the intuition between GSC and GSS.

Versions of this general theory have been applied broadly and spawned new lines of research in different fields, including networks, global games, political science, constrained optimization in economic theory, and dynamic games, among others.

A range of applications of monotone games in the field of networks is available in Goyal (2007), Jackson (2008), Bramoullé et al. (2014), and Jackson and Zenou (2014), among others.

Applications in the field of global games are available in Carlsson and van Damme (1993), Morris and Shin (2003), Frankel et al. (2003), Hoffmann and Sabarwal (2019a), and Hoffmann and Sabarwal (2019b), among others.

Applications in political science are highlighted in Ashworth and de Mesquita (2005) and Shadmehr and Bernhardt (2011), among others.

Developments in the field of directional constrained optimization and economic theory are available in Quah (2007), Quah and Strulovici (2009), and Barthel and Sabarwal (2018), among others.

Applications to dynamic games are available in Curtat (1996), Amir (1996), Echenique (2004), Balbus, Reffett, and Woźny (2014), and Feng and Sabarwal (2020), among others.

This is a partial list and many more applications are available.

Existing applications focus more on games with strategic complements, increasing maximizers, aggregative games, or on special cases such as affine best responses. Applications with strategic substitutes or with complements and substitutes together are less developed.

A unified treatment of foundations of the theory of monotone games makes this material more accessible to a broader audience and increases the scope of its applications. As applications proliferate, in turn, their particular patterns inform development of theory. The core principles studied here emerge in a large body of human and socioeconomic interaction with interdependent effects, and therefore, there is a large potential benefit of an increase in our shared understanding of these ideas.

References

Acemoglu, D., & Jensen, M. K. (2013). Aggregate Comparative Statics. *Games and Economic Behavior, 81*, 27–49.

Amir, R. (1996). Continuous Stochastic Games of Capital Accumulation with Convex Transitions. *Games and Economic Behavior, 15*, 111–131.

Ashworth, S., & de Mesquita, E. B. (2005). Monotone Comparative Statics for Models of Politics. *American Journal of Political Science, 50*(1), 214–231.

Balbus, Ł., Reffett, K., & Woźny, Ł. (2014). A Constructive Study of Markov Equilibria in Stochastic Games with Strategic Complementarities. *Journal of Economic Theory, 150*, 815–840.

Barthel, A.-C., & Hoffmann, E. (2019). Rationalizability and Learning in Games with Strategic Heterogeneity. *Economic Theory, 67*, 565–587.

Barthel, A.-C., & Sabarwal, T. (2018). Directional Monotone Comparative Statics. *Economic Theory, 66*, 557–591.

Bramoullé, Y., Kranton, R., & D'Amours, M. (2014). Strategic Interaction and Networks. *American Economic Review, 104*(3), 898–930.

Bulow, J. I., Geanakoplos, J. D., & Klemperer, P. D. (1985). Multimarket Oligopoly: Strategic Substitutes and Complements. *Journal of Political Economy, 93*(3), 488–511.

Carlsson, H., & van Damme, E. (1993). Global Games and Equilibrium Selection. *Econometrica, 61*(5), 989–1018.

Curtat, L. (1996). Markov Equilibria of Stochastic Games with Complementarities. *Games and Economic Behavior, 17*, 177–199.

Dubey, P., Haimanko, O., & Zapechelnyuk, A. (2006). Strategic Complements and Substitutes, and Potential Games. *Games and Economic Behavior, 54*, 77–94.

Echenique, F. (2004). Extensive-Form Games and Strategic Complementarities. *Games and Economic Behavior, 46*(2), 348–364.

Feng, Y., & Sabarwal, T. (2020). Strategic Complements in Two Stage, 2x2 Games. *Journal of Economic Theory*, 190, forthcoming.

Frankel, D. M., Morris, S., & Pauzner, A. (2003). Equilibrium Selection in Global Games with Strategic Complementarities. *Journal of Economic Theory, 108*(1), 1–44.

Fudenberg, D., & Tirole, J. (1984). The Fat-Cat Effect, the Puppy-Dog Ploy, and the Lean and Hungry Look. *American Economic Review, 74*(2), 361–366.

Goyal, S. (2007): *Connections: An Introduction to the Economics of Networks*. Princeton University Press.

Hoffmann, E. J., & Sabarwal, T. (2019a). Equilibrium Existence in Global Games with General Payoff Structures. *Economic Theory Bulletin*, *7*, 105–115.

Hoffmann, E. J., & Sabarwal, T. (2019b). Global Games with Strategic Complements and Substitutes. *Games and Economic Behavior*, *118*, 72–93.

Jackson, M. O. (2008). *Social and Economic Networks.* Princeton University Press.

Jackson, M. O., and Y. Zenou (2014): "Games on Networks," in *Handbook of Game Theory)*, ed. by P. Young, and S. Zamir, vol. 4, chap. 3, pp. 95–164. Elsevier Science.

Jensen, M. K. (2010). Aggregative Games and Best-Reply Potentials. *Economic Theory*, *43*(1), 45–66.

Monaco, A., & Sabarwal, T. (2016). Games with Strategic Complements and Substitutes. *Economic Theory*, *62*(1), 65–91.

Monderer, D., & Shapley, L. (1996). Potential Games. *Games and Economic Behavior*, *14*, 124–143.

Morris, S., and H. S. Shin (2003): "Global Games: Theory and Applications," in *Advances in Economics and Econometrics (Proceedings of the Eighth World Congress of the Econometric Society)*, ed. by M. Dewatripont, L. Hansen, and S. Turnovsky, chap. 3, pp. 56–114. Cambridge University Press, Cambridge.

Quah, J. K.-H. (2007). The Comparative Statics of Constrained Optimization Problems. *Econometrica*, *75*(2), 401–431.

Quah, J. K.-H., & Strulovici, B. (2009). Comparative Statics, Informativeness, and the Interval Dominance Order. *Econometrica*, *77*(6), 1949–1992.

Rosenthal, R. W. (1973). A Class of Games Possessing Pure-Strategy Nash Equilibria. *International Journal of Game Theory*, *2*(1), 65–67.

Roy, S., & Sabarwal, T. (2008). On the (Non-)Lattice Structure of the Equilibrium Set in Games With Strategic Substitutes. *Economic Theory*, *37*(1), 161–169.

Roy, S., & Sabarwal, T. (2010). Monotone Comparative Statics for Games with Strategic Substitutes. *Journal of Mathematical Economics*, *46*(5), 793–806.

Roy, S., & Sabarwal, T. (2012). Characterizing Stability Properties in Games with Strategic Substitutes. *Games and Economic Behavior*, *75*(1), 337–353.

Shadmehr, M., & Bernhardt, D. (2011). Collective Action with Uncertain Payoffs: Coordination, Public Signals, and Punishment Dilemmas. *American Political Science Review*, *105*, 829–851.

Index

T. Sabarwal, *Monotone Games*,
https://doi.org/10.1007/978-3-030-45513-2

Printed in the United States
by Baker & Taylor Publisher Services